This is a continuation in the series of publications produced by the Center for Advanced Concepts and Technology (ACT), which was created as a "skunk works" with funding provided by the CCRP under the auspices of the Assistant Secretary of Defense (NII). This program has demonstrated the importance of having a research program focused on the national security implications of the Information Age. It develops the theoretical foundations to provide DoD with information superiority and highlights the importance of active outreach and dissemination initiatives designed to acquaint senior military personnel and civilians with these emerging issues. The CCRP Publication Series is a key element of this effort.

Check our Web site for the latest CCRP activities and publications.

www.dodccrp.org

DoD Command and Control Research Program

Assistant Secretary of Defense (NII)
&
Chief Information Officer
Mr. John G. Grimes

Principal Deputy Assistant Secretary of Defense (NII)
Dr. Linton Wells, II

Special Assistant to the ASD(NII)
&
Director of Research
Dr. David S. Alberts

Library of Congress Cataloging-in-Publication Data

Alberts, David S. (David Stephen), 1942-
Campaigns of experimentation : pathways to innovation and transformation / David S. Alberts, Richard E. Hayes.
  p. cm. -- (Information age transformation series)
Includes bibliographical references.
ISBN 1-893723-15-1
1. Military research--United States--Methodology. I. Hayes, Richard E., 1942- II. Code of best practice for experimentation. III. Title. IV. Series.

U393.A66697 2005
355'.07'0973--dc22
                                                    2004030160

March 2005
February 2006

Information Age Transformation Series

CODE OF BEST PRACTICE

CAMPAIGNS
of
EXPERIMENTATION

Pathways to
Innovation and Transformation

David S. Alberts
Richard E. Hayes

Table of Contents

List of Figures

ACKNOWLEDGMENTS

This Code of Best Practice was undertaken at the specific request of Major General James M. Dubik (USA), then the J9 at Joint Forces Command. In that role, he was responsible for Joint experimentation, one of the most important elements of DoD transformation efforts and clearly on the critical path to the future capabilities that the U.S. military needs for the twenty-first century. MG Dubik kept a well-worn copy of the original *Code of Best Practice for Experimentation* in his office (and often with him as he moved around the organization) and told us that it had been genuinely helpful. However, he recognized that individual experiments had limited value and that he, and other key players in his command and other DoD organizations, required a better understanding of how campaigns of experimentation could be conceived and executed—campaigns that would build knowledge and develop capabilities. He therefore requested that the CCRP gather ideas from the best available people and develop a new COBP that would extend the original chapter on campaigns in the *Code of Best Practice for Experimentation*.

We found that responding to this challenge was not trivial. While the basics for designing individual experiments are widely understood and practiced throughout the scientific community, deliberately crafted campaigns of experimentation are much less common. Even less experience and expertise exist in organizing and executing campaigns of experimentation in the complex arena of transformation. Hence, this book should not be considered definitive, but rather an effort to offer ideas and lessons learned over the past decade from working with DoD, coalition partners, and interagency efforts to coevolve disruptive innovations, particularly in the command and control arena.

We were fortunate, therefore, to be able to draw on the insights and experience of many colleagues. Dr. Daniel T. Maxwell contributed valuable early thinking on the topic. Dr. Richard Kass and Mr. Shane Deichman provided thoughtful input from their perspectives within Joint Forces Command. Drs. Dennis K. Leedom and David F. Noble also contributed thoughtful ideas. The members of the Information Age Metrics Working Group (IAMWG), senior personnel who assemble monthly to look at important issues, gave us rich feedback on the early draft material and participated in collegial discussions of key topics covered in the book. Regular members of the group include Dr. Ed Smith, John Poirier, Dennis Popiela, Dr. Mike Bell, Mark R. Sinclair, Dr. Mark Mandeles, Julia Loughran, Kirsch Jones, Eugene Visco, Dr. Larry Wiener, Manual Miranda, Pat Curry, Donald G. Owen, Mitzi Wertheim, RADM Evelyn J. Fields (ret.), Dr. Paul Hiniker, and Dr. David T. Signori. Four members of the IAMWG, Dr. Larry Wiener, Dr. Mike Bell, John Poirier, and Dr. Mark Mandeles, were particularly generous with their time, developing the Checklist for Individual Experiments included in the Appendix to this volume, as well as thoughtful feedback on draft materials. Their checklist effort built on a Checklist for Individual Experiments in the Intelligence Community created by Dr. Annette Krygiel, based on her experience and the original *COBP for Experimentation*.

We were very ably supported in drafting this book by Jonathan Tarter and Eugene Hopkins, who conducted bibliographic research and performed fact checking as required. Joseph Lewis created the graphics and cover artwork, acted as the technical editor, and did the key layout work. In those processes, he made valuable suggestions that helped us make the text clearer and better organized. Margita Rushing handled the process of moving the draft through the publication process.

Dr. David S. Alberts   Dr. Richard E. Hayes

PREFACE

This book, the seventh in the CCRP's series on Information Age Transformation, addresses a critical core competency for a twenty-first century military, or for that matter any organization that needs to embrace disruptive innovation to survive. This core competency is the ability to successfully undertake campaigns of experimentation designed to result in disruptive innovation.[1] The introduction[2] of the first book in this series, *Information Age Transformation: Getting to a 21st Century Military*, acknowledges the fundamental obstacle to progress: "Military organizations are, by their very nature, resistant to change," and the reason why "this is, in no small part, due to the fact that the cost of error is exceedingly high." But change we must, despite the formidable challenges that lie ahead.[3]

Experimentation is both an opportunity to explore "outside the box" and a proven method of risk management. When properly conceived and executed, campaigns of experimentation strike the proper balance between innovation and risk. As a result, organizations are able to embrace new concepts, organi-

---

[1] This call for a new core competency was first voiced in the preface of the second book in this series on Information Age Transformation, the *Code of Best Practice for Experimentation*, p. xi.

[2] Alberts, *Information Age Transformation*. p. 1.

[3] Ibid. Chapter 4: Dealing with the Challenges of Change. pp. 25-29.

zational forms, approaches to command and control, processes, and technologies. In other words, they are able to accomplish disruptive (transformational) change with an acceptable level of risk. Given the nature of military institutions, achieving the proper balance is not likely to occur without developing a broad-based understanding of, and a significantly improved ability to conduct, campaigns of experimentation.

It is hoped that this book, together with its companion volumes in the CCRP's Information Age Transformation Series, will contribute to this better understanding and improve the state of the practice.

Dr. David S. Alberts

Director of Research, OASD(NII)

CHAPTER 1

INTRODUCTION

The Department of Defense (DoD), like other institutions, businesses, and organizations, is engaged in a transformation that is, in effect, an adaptation to the Information Age. In the case of DoD, this transformation is a response to a significant change in the nature of the missions that it must execute, as well as an opportunity to become more effective by adopting Information Age concepts and technologies.

While there is general agreement regarding the general nature of the changes needed, the details of DoD transformation remain a journey into the unknown. In other words, we may agree that moving to a more network-centric organization is required, but the specifics in terms of new approaches to command and control, organization, doctrine, processes, education, training, and even the capabilities we require have yet to be developed adequately. Moreover, substantial issues must be addressed before the necessary concepts can be developed, articulated, and assessed. If DoD's Information Age transformation is to be successful, it needs to be informed by a coherent set of lessons learned, experiments, and empirical analyses. This will provide the feedback necessary to keep the effort on track.

PURPOSE AND NATURE OF EXPERIMENTATION

Experimentation plays a vital role in transformation. Experimentation contributes to and thus advances a body of knowledge that, when applied, allows us to develop new capabilities. While some think of experimentation as simply conducting individual experiments, experimentation is a *process* rather than just a collection of experiments—a process that (1) combines and structures experimental results much in the way that individual bricks are fashioned into a structure for a purpose, and (2) steers future experimentation activities.

Individual experiments, however ambitious they might be, are limited in what they can explore and thus conclude. There is a limit to the number of variables and the relationships among them that can be reasonably explored or tested in a single experiment. Further, the results achieved in one experiment need to be replicated under the same or similar conditions in other experiments before they can be considered reliable and valid. Moreover, the specific set of assumptions and conditions that defines the environment of an experiment needs to be fully explored if the results are to be properly interpreted, extrapolated, and understood. Hence, the potential contribution of any one experiment is limited.

The value of a given experiment depends upon what is known and what other experiments have been or could be conducted. Hence, the value of a given experiment depends as much (if not more) on the process of knowledge development as the knowledge that it is able to generate by itself. That is, the value of a single piece of a puzzle is greatly enhanced when we have or can get more of the pieces. The value proposition for experiments can be thought of as analogous to the value proposition for networks, in which the value of the network goes up exponentially with the number of nodes.[4] Hence, transformational experiments should be part of a well-designed *series* of experiments and related activities that we call a *campaign.*

Thinking of experiments in the context of campaigns is important not only because it greatly enhances the value of the experiments that will be undertaken, but also because it helps avoid one of the most common pitfalls associated with the design and conduct of individual experiments: the tendency to push a given experiment beyond its feasible limits. Trying to do too much in a single experiment greatly increases the probability that the experiment will generate little in the way of useful information or experience, and therefore be of little value. Overly ambitious efforts are a waste of time, talent, and resources.

TRANSFORMATION AND EXPERIMENTATION

To be better able to explore and understand the complex issues involved in military transformation, *campaigns of experimentation*, each consisting of a set of experiments, complementary analyses, and synthesis activities, need to be conceived, orchestrated, and harvested. Each campaign needs to be focused on a specific set of issues or capabilities. Each will need to include a number of well-chosen and properly sequenced experiments, some of which cannot be fully designed until the results of other, earlier experiments have been conducted and analyzed. It is also good practice when scoping any given experiment or analysis to have in mind the nature of future experiments that might be conducted.

---

[4] Alberts et al., *Network Centric Warfare.* p. 245ff.

DoD will need to undertake a fairly large number of campaigns of experimentation because, like individual experiments, individual campaigns have a limited ability to explore the behavior of a set of variables across the broad range of contexts relevant to DoD.

The Information Age transformation of DoD requires that we greatly expand our understanding of the tenets of Network Centric Warfare (NCW),[5] the principles associated with Power to the Edge,[6] and the application of these principles to the design, development, and deployment of mission capability packages (MCPs).[7] An enhanced understanding of what constitutes sufficient *shared awareness*—that level of shared awareness necessary to support self-synchronization—is needed so that we may assess and understand the implications of different approaches to command and control, organizations, and processes. These interrelated issues will require several campaigns of experimentation to explore them fully.

NATURE OF THE CAMPAIGNS REQUIRED

Three types of campaigns of experimentation are needed to fully support an Information Age transformation. These include campaigns that:

1. Explore the tenets of NCW;

2. Coevolve MCPs; and

3. Explore coalition and interagency operations.

---

[5] Alberts, *Information Age Transformation*. p. 7.

[6] Alberts and Hayes. *Power to the Edge.*

[7] Alberts, *Information Age Transformation*. p. 74.

EXPLORING THE TENETS OF NCW

The tenets of Network Centric Warfare[8] form the theoretical foundation for developing transformational mission capability packages:

1. A robustly networked force improves information sharing.

2. Information sharing and collaboration enhance the quality of information and shared situational awareness.

3. Shared situational awareness enables self-synchronization.

4. These, in turn, dramatically increase mission effectiveness.

However, these tenets provide only a general direction and set of considerations for those designing and implementing mission capability packages.

There is a growing body of evidence supporting these tenets,[9] but the available evidence involves applications that barely

---

[8] Alberts et al., *Network Centric Warfare*. pp. 193-197.

[9] University of Arizona. Study: Decision Support for U.S. Navy's Combined Task Force 50 during Enduring Freedom.
Reinforce. Study: Multinational Operations (During IRTF (L) trial of AMF (L); Amber Fox; and ISAF 3).
RAND. Study: Stryker Brigade Combat Team.
PA Consulting Group. Study: Joint U.S./U.K. Combat Operations in Operation Iraqi Freedom.
SAIC. Study: Air to Ground Operations in DCX (Phase 1), Enduring Freedom and Iraqi Freedom.
Booz Allen Hamilton. Study: NCO Conceptual Framework: Special Operations Forces Case Study.

scratch the surface of what is possible and reflect only a limited set of conditions. Thus, while the evidence generally supports the tenets, a rich and full understanding has yet to emerge. To more fully explore these tenets, two types of campaigns would be useful. The first would be a set of campaigns focused on the nature of shared awareness and the conditions that make it possible. These campaigns would help us determine the nature of the information processing capabilities, processes, and conditions that are required to achieve widespread sharing of information, productive collaboration, quality awareness, and the development of shared awareness. The second set of campaigns would start with various levels of shared awareness and explore the nature of the organizations, doctrines, cultures, and approaches to command and control that are best able to leverage shared awareness.

COEVOLUTION OF MISSION CAPABILITY PACKAGES

The region (of the knowledge landscape) to be explored in an experiment (or campaign) is defined by the dimensions of a mission capability package. Constraining the values for one of these dimensions serves to place a section of the region off limits. Thus, the solutions considered are drawn from a subset of the information and means that exist. It is like being forced to search with one or more senses taken away.

Figure 1 graphically depicts the effect that ignoring or constraining selected elements of a mission capability package has on the set of solutions that can be examined. Because we do not know where the better or best solutions are, such constraints serve to limit the value of the solutions found. If we place constraints on any element of a MCP, they narrow the field of study for that element. If we continue to limit and con-

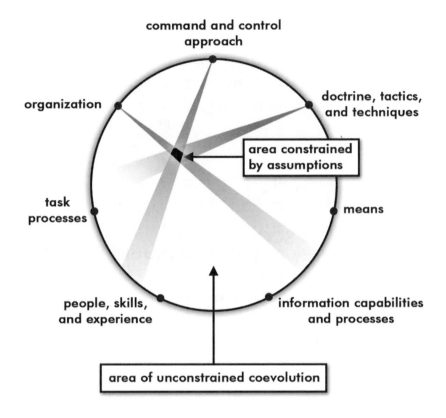

Figure 1. Mission Capability Package Landscape

strain each element, then the area of exploration that remains open to us will be reduced to a very small region where our constraints intersect. For example, if we limit our study to a narrow set of assumptions about (1) organization, (2) command and control approaches, and (3) doctrine, tactics, and techniques, then only the small black area of Figure 1 will remain. When compared to the whole unconstrained area of the MCP landscape, this region of constrained exploration constitutes only a very small fraction of our true options.

COALITION AND INTERAGENCY NCW

The ability to assemble a coalition and effectively leverage the capabilities of coalition partners is fundamental to success in the twenty-first century. A growing number of countries have indicated that they believe in the promise of network-centric operations (NCO).[10] Given the nature of both NCO and coalitions, it is important that multilateral research and experimentation be undertaken. One country's solution, no matter how well it works for that country, is unlikely to work well for or be accepted by others.

PURPOSE AND SCOPE OF THIS CODE OF BEST PRACTICE (COBP)

This book is a logical follow-on to the *Code of Best Practice for Experimentation*,[11] which was focused on the design and conduct of individual experiments. That book introduced the idea of a campaign of experimentation and compared and contrasted the differences between an individual experiment and a campaign of experimentation.[12]

The ability to design and conduct individual experiments constitutes a necessary—but not sufficient—core capability to conceive, design, and conduct successful campaigns of experimentation. Thus, the purpose of this book is to build upon the

---

[10] U.K. Ministry of Defense. "Network Enabled Capability." 2004.
   Swedish Armed Forces Headquarters. "Network Based Defenset." 2002.
   Wik, "Networked-Based Defense for Sweden." 2002.
   "Strategic Vision: A Paper by NATO's Strategic Commanders." 2004.
   Fourges. "Command in Network-Centric War." 2001.
   Kruzins. "Factors for Network Centric Warfare." 2002.

[11] Alberts et al., *Code of Best Practice for Experimentation*. 2002.

[12] *Ibid.*, p. 44.

discussions in the earlier *COBP for Experimentation* and explain in greater detail the nature of a campaign of experiments, how these campaigns should be effectively conceived, designed, and executed, and how their results can be harvested in the context of the ongoing transformation of military organizations and operations.

While discussions of campaigns of experimentation have broad applications, this book focuses on *transformational* campaigns that seek breakthroughs in knowledge or capability, rather than those designed to marginally improve or refine our knowledge or a given capability. In the vernacular of innovation,[13] this book focuses on the role and conduct of campaigns of experimentation that involve *disruptive* innovation, rather than *sustaining* innovation. These campaigns are inherently more complex and involve greater risk than those focused on sustaining innovation, but they have correspondingly greater potential. In contrast to the incremental improvements in understanding and capabilities that result from exercises or systems tests, these campaigns share the more complete vision of research long associated with DARPA (Defense Advanced Research Projects Agency).[14]

JFCOM (Joint Forces Command) recognizes this distinction. Figure 2 is their graphic depiction of the relationship between sustaining and disruptive innovation.[15]

---

[13] Innovation: the act of introducing something new. *American Heritage Dictionary.*

[14] "DARPA's work is high-risk and high-payoff precisely because it fills the gap between fundamental discoveries and their military uses."
DARPA Strategic Plan. p. 4.

[15] Dubik, "Campaign Plan 2003-2009." 2004.

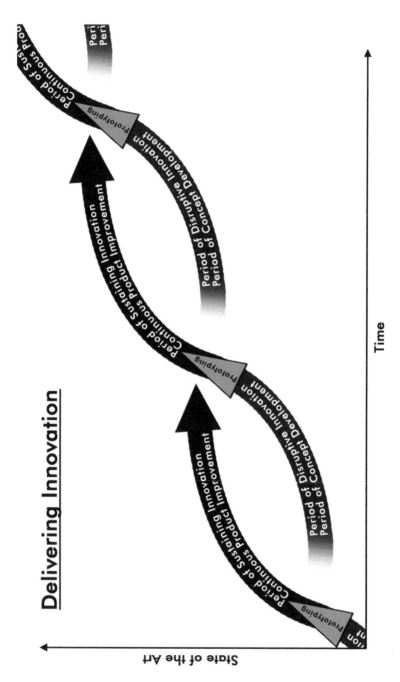

Figure 2. Sustaining Innovation v. Disruptive Innovation

Their concept-based campaigns of experimentation focus on disruptive innovation rather than on sustaining innovation.

INTENDED AUDIENCE

This COBP is intended for several audiences:

1. A specific audience, those who are directly engaged in experimentation activities, whether they are focused on identifying and developing new concepts, evaluating the growing body of evidence regarding Network Centric Warfare/Operations and making related decisions regarding the development and adoption of new approaches and capabilities, or those who are directly involved in planning and conducting experiments;

2. A military audience, those interested in military transformation; and

3. A general audience, those who are interested in Information Age transformation regardless of organization or domain.

Given the need to conduct network-centric warfare/operations with a variety of military, government, and non-governmental organizations, campaigns of experimentation as well as individual experiments will need to include participants from other government agencies and different countries, including those who speak different languages, have different areas of expertise, and who could have very different goals. Transformation and twenty-first century missions take us beyond Joint operations into operations that involve military coalitions, interagency partnerships, and public and private organizations. Success demands that all of the participants understand and collaboratively contribute to the formulation and execution of campaigns of experimentation. This book is intended to provide a common frame of reference for these diverse audiences in the hope that it will enable them to collaborate more effectively on experimentation campaigns, as well as to design and conduct their own experimentation activities in a way that maximizes the potential of the data and results generated to build upon one another.

ORGANIZATION OF THE COBP

OVERVIEW

This volume begins with a discussion of the nature of experimentation, DoD transformation, and innovation. It continues with an explanation of the nature of a campaign of experimentation, details regarding the planning and execution of such campaigns, and a review of the current state of the practice. The book concludes with a discussion of harvesting results across campaigns and a look at the way ahead.

CHAPTER BY CHAPTER

- Chapter Two provides an overview of the scientific nature and history of experimentation to properly set the stage for a discussion of how we must approach experimentation today.

- Chapter Three introduces the subject of transformation and discusses why DoD must weigh the risks of revolutionary change against the risks of business as usual. A transformed DoD will not only require new tools and doctrine, but new attitudes towards the concepts of change, risk, cost, and success.

- Chapter Four examines the differences between transformation and modernization, and more specifically, the differences between disruptive and sustaining innovation. Detailed comparisons to business and industry demonstrate the methods and mindsets needed to survive, grow, and succeed in the long term.

- Chapter Five defines and explores the key concept of a campaign of experimentation. This discussion touches on the three primary types of experiments, their uses, and the various issues that arise when conducting each. The second half of this chapter focuses on the key issue of maintaining fidelity and control throughout the conduct of a campaign of experimentation.

- Chapter Six presents the various aspects and phases of a properly conducted campaign of experimentation and describes many of the issues that must be taken into account from the beginning of the campaign, as well as managed throughout the entire process.

- Chapter Seven offers a look into a number of examples of campaigns of experimentation from the last several years.

- Chapter Eight discusses some of the larger, overarching challenges associated with collecting and utilizing the information products of a campaign of experimentation.

- Chapter Nine concludes the book with a final look at the issues that we must continue to confront and overcome in the development and execution of campaigns of experimentation. This chapter also includes several examples of important campaign concepts and how those campaigns might be carried out.

CHAPTER 2

EXPERIMENTATION

SOURCES OF KNOWLEDGE

*E*xperiment comes from the Latin word *experiri* meaning "to try." The Latin word *experimentum* means a trial or a test. In modern English, the verb *to experiment* is commonly defined as "any action or process undertaken to discover something not yet known or to demonstrate something known."[16] Thus the word is used to refer both to formal behaviors that are rigorous (i.e., behaviors that adhere to a set of scientific principles) as well as to refer to informal behaviors that are expressions of curiosity.

It was not that long ago (as late as the sixteenth century) when observation and experimentation became established in the Western world as the primary means by which we advanced knowledge and our understanding of the world. Prior to that time, truth and knowledge had been thought to be solely a product of intuition or reason. With this shift in the recognized source of knowledge came the question of "How do we know what is a fact or truth?"

---

[16] *Webster's New World Dictionary.* 1995.

Descartes[17] believed that you knew what was true when all reasons for doubt were removed. Bacon[18] suggested an analytical approach. In England, a Royal Society[19] was formed to document and certify observations and methods. This was the beginning of the widespread acceptance and application of an empirically based approach to advancing understanding. Systematic observation, experimentation, and analysis now form the core of satisfying our collective curiosity and answering specific questions about "Why?" and "How?"

Over time, what is now known as the scientific method was developed. This accumulation of lessons learned and theory provides guidance on the design of experiments, the collection of data, its analysis, and the nature of the conclusions that can be drawn. Because any real-world observations that we make constitute a very small sample of reality under some limited conditions, statistical theory, methods, and tools play a major role in the process of knowledge acquisition.

LEAPS: GREAT AND SMALL

The written history of science tends to jump from breakthrough to breakthrough and not dwell on all of the hard work that is required to "set up" breakthroughs or the difficult tasks involved in filling in the details—learning the implications and limitations of the broad breakthroughs. Rarely is a breakthrough solely the product of one individual's efforts. They are more typically the results of series of convergences of ideas, things, and people that often occur over long periods of time. Breakthroughs by definition involve discontinuities and new

---

[17] Routledge Encyclopedia of Philosophy. 2004.

[18] The Internet Encyclopedia of Philosophy. 2004.

[19] The Royal Society of London. 2004.

ideas or constructs. They often involve drawing upon widely different fields of study or application.[20] There is no mystery here. The insights that are possible within the knowledge, tools, and practices of a given field will naturally be limited. Insights that require a new perspective, a missing fact, or a different tool or process will remain undiscovered until someone or some team brings everything together.

Throughout most of history, most scientific advances were the results of unplanned convergences, naturally occurring as a function of the ability to share information and the interests of independent parties, with serendipity playing a major role (at least as far as timing is concerned). Thus, the march of scientific advances that appears orderly in hindsight is not an organized march at all, but almost a random walk, meandering this way and that through both time and space.

James Burke, author of the popular book *Connections*,[21] created a popular television mini-series that recreated a number of these "walks." For example, the creation of the carburetor by the nineteenth-century German engineer Wilhelm Maybach involved the confluence of gasoline and a perfume sprayer. Or in the case of a new approach for directing anti-aircraft batteries, the early twentieth-century American mathematician Norbert Wiener put the physiological concept of homeostasis together with mathematical algorithms to produce a feedback mechanism that eventually became known as cybernetics.[22]

---

[20] The interdisciplinary team concept in Operations Research is a recognition of the need for variety—variety of skills, perspectives, expertise, and experience.

[21] Burke, *Connections*. 1978.

[22] Burke provides examples of convergence in an article in *Forbes*: "Now What?" October 4, 1999.

The adoption of the results of these breakthroughs is usually the subject of another, unrelated series of convergences. The same process applies to the resulting coevolution that takes place as social and organizational adaptations occur in response to new capabilities in far flung domains. Adaptations are often paced by the frequency and success of efforts that (1) serve to explore the specific conditions under which these new capabilities have value, and (2) refine the capability so that it can be usefully applied.

Thus, great leaps serve to open up new fields of investigation and establish pathways of innovation, while small leaps pave the paths so that others may easily follow in the footsteps of the pioneers and first adopters. These unplanned advances occur as a function of many unrelated events. What advances occur and the frequency of their occurrence are emergent properties of the larger "system" within which they occur. They are a function of the nature and level of education, research funding, priorities, and scientific and communications infrastructure, as well as many other factors.

PROCESS OF EXPERIMENTATION

Experimentation is, as explained above, a process that may or may not be "orchestrated." Experimentation involves not just the conduct of experiments but also the following elemental tasks of knowledge acquisition: development of a theory, construction of a conceptual model that embodies the key elements of the theory, formulation of questions (Descartes' doubts) and hypotheses (Bacon's analytical foci), collection of evidence, and analysis. This process is empirically based. Hence, progress depends upon an accumulation of data and analytical findings. Thus each of these elemental tasks needs to be undertaken in a way that at least permits, if not actively supports and encourages, reuse. If this is not done, the building blocks necessary to add to our body of knowledge will not be created, arranged, and rearranged to create a foundation that supports new layers of understanding. As a consequence, the process of creating knowledge will not proceed with any deliberate speed. *Reuse* here applies to ideas, information about investigations conducted, data collected, analyses performed, and tools developed and applied. In terms of experiments, it implies *replication*. Reuse, and hence progress, is maximized when attention is paid to the principles of science that prescribe how these activities should be conducted, how peer reviews should be executed, and when attention should be paid to the widespread dissemination of findings and conclusions.

An analogy can be drawn between the relationships between platforms and the network in NCW and the relationships between an experiment and a campaign of experimentation. In the case of NCW, it is really all about the network, that is, the interactions among the entities—what takes place in the social/organization domain. As a result, the network is the biggest source of combat power.[23] The value of any single platform is relatively small compared to the value derived from networking several of them together. The same is true of a single experiment. It is the synergy that one gets from the collective experiences and the interactions that propel science, the development of new concepts, and their applications.

EMERGENT V. MANAGED EXPERIMENTATION

Until recently, the pace of naturally occurring invention and scientific progress seemed, to many, to be quite satisfactory. In

---

[23] Network Centric Warfare Department of Defense Report to Congress. 2001.

fact, for some the accelerating pace of change that became discernable as the Industrial Revolution matured was (and has continued to be) disturbing.[24]

Pressures for change come from both external and internal sources. The adage, "necessity is the mother of invention," speaks to external pressures to solve a particularly urgent problem. Curiosity and a belief that things could be better or better understood also contribute to a desire to accelerate the natural pace of scientific advance and accelerate the development of new ways of doing things.

Given the major security challenges facing the Western world in general and the United States in particular, there is an urgent need to bring the power of scientific thought to bear on both understanding the threat and developing more effective responses to that threat. Hence, experimentation, as the method for developing new understandings and improved applications, is being increasingly employed to explore new approaches, such as NCO. However, the perceived urgency must not be allowed to reduce the quality of the experimentation process or the quality of the new knowledge and applications. The purpose of campaigns of experimentation is to ensure that these efforts are organized for success.

---

[24] From 1811 to 1817, unemployed laborers launched a movement called Luddism (named for leader Ned Ludd) in England to protest violently against the use of machines in factories, which reduced the number of jobs for manual laborers. Thousands of desperate and starving men attacked various manufacturing centers all over England, but the movement was finally ended through military intervention. Because of these events, individuals today who oppose technological progress are sometimes referred to as Luddites. See: Spartacus Educational. "The Luddites."

EXPERIMENTATION RISKS AND REMEDIES

The events of 9/11 have resulted in a desire to leverage the concepts and capabilities of the Information Age as soon as possible. To accomplish this, we are undertaking sharply focused, single-threaded (streamlined) campaigns of experimentation. This approach, although understandable in light of the resources currently allocated, entails of a number of significant risks that need to be recognized and managed. These risks include (1) moving ahead without sufficient evidence and understanding, (2) prematurely settling on an approach, (3) confining explorations to the Industrial-Information Age border, (4) progressing by trial and error as opposed to being guided by theory, and (5) failing to capitalize on the creativity present in the force.

PROCEEDING WITH INSUFFICIENT EVIDENCE

The first risk identified is proceeding without sufficient understanding or evidence. Clearly the urgency of the task at hand makes it foolish to insist on full academic research standards before adopting any meaningful conclusions from experimentation. On the other hand, it would be just as foolish and potentially more costly in lives and resources to rely on a single or small number of experiments to support important decisions. This statement would seem to be rather obvious, but we continue to encounter briefings and reports that urge major decisions based on individual experiments, arguing that they have provided "proof" that a key concept or idea has military value.

This is a very naïve and dangerous practice for several reasons. First, without sufficient opportunity for replication, it is impossible to tell whether a result is really attributable to a new concept or not. Second, it is impossible to know with any degree of confidence that the new concept is robust (would work under a range of circumstances). Part of robustness is how the concept would fare when an adversary has an opportunity to adapt. There are a number of ways to design a campaign of experimentation to mitigate the risks associated with adopting approaches or systems that have not been adequately explored.

This, of course, assumes that the experimentation activities being conducted are well-conceived and are implemented in a way that maximizes the generation of useful information. This, unfortunately, is not always the case. Although DoD, as an institution, is getting better at conducting experimentation events, these are all too often flawed, sometimes fatally because of problems with experiment design, data collection, and analysis. There are obvious and recurring problems arising from choosing to collect what is easily collected and substituting opinion for empirical data and rigorous analysis.

The *COBP for Experimentation* identifies the important considerations involved in conducting experiments, discusses the consequences of failing to adhere to best practices, and provides some examples of what not to do. For the purposes of this discussion, we assume that individual experimentation activities are undertaken according to something that approaches best practice and confine ourselves to a discussion of campaigns. Of paramount concern is proper instrumentation so that empirical evidence (as opposed to just opinions or anecdotes) can be collected.

The most direct approach is to increase the number and variety of experimentation activities, conducting them in parallel if necessary, to generate more data and experience. To maximize the ability of a number of different experimentation activities to contribute to a coherent body of evidence, a conceptual framework that identifies the variables and the relationships among them that are thought to be important must exist. This conceptual framework also needs to provide operational definitions for the variables of interest so that their values can be measured. It is equally important that a set of instruments or tools be available to ensure that the measurements taken are comparable across experimentation activities.

The Office of the Assistant Secretary of Defense for Networks and Information Integration (OASD[NII]) and the Office of Force Transformation (OFT) within OSD continue to collaborate on a Network Centric Operations Conceptual Framework[25] and a set of case studies. The NCO Conceptual Framework provides a set of measures thought to be important to understand and document the application of NCW theory to military operations. The case studies illustrate how these measures can be applied to real-world situations. There is also a NATO effort[26] with participation from selected non-NATO countries underway that is building a Conceptual Model designed to assess network-centric approaches to command and control, one that can be used to understand and assess operations and investigate the tenets of NCW. Finally, OASD(NII) is also cooperating with OASD(HLD) to extend these frameworks to the heavily

---

[25] Office of Force Transformation. "NCO Conceptual Framework Version 1.0."

[26] The NATO Research and Technology Office sponsors several panels including the SAS (Studies, Analysis, and Simulation) Panel. The SAS Panel sponsors research groups including one under the chairmanship of the CCRP that is focused on exploring new C2 concepts. The effort features the development of a conceptual model for C2.

interagency and civilian realm of Homeland Defense. These efforts are producing products that provide a growing list of variables that can be used to help understand transformation. They provide a point of departure for the design of the data collection and analysis efforts central to the conduct of coherent campaigns of experimentation.

PREMATURE NARROWING OF FOCUS

In an effort to get something to the field as quickly as possible, there is currently a tendency to take a promising new approach and "fast track" it. That is, to focus efforts on bringing an idea, concept, or technology quickly (and potentially prematurely) to "market." Moving immediately from an idea to heavily controlled experimentation environments or richly realistic ones, without due attention to discovery, is likely to inhibit or prevent open-ended inquiry and adversely affect the likelihood of developing counterintuitive insights. This failure to provide ample opportunity for discovery and/or attention to appropriately maturing concepts or knowledge can have potentially serious consequences. What is the opportunity cost associated with restricting exploration of the possibilities? Settling on a promising idea, while ignoring a far better idea, even though the former may represent a step forward, means foregoing significant benefits for some period of time and is not consistent with the goal of transformation.

Many DoD organizations' plans call for one or at most a small number of discovery-oriented activities. These discovery activities are normally conducted sequentially, and as a result the pressures (in the case of the near-term campaign) to get something to the field quickly result in the selection of the "best of breed," which often translates into the most mature instantiation, that is, often an off-the-shelf technology or system rather than a fully coevolved approach with technology that is well-suited to the concept. In fact, this tendency to equate *approach* or *capability* with a specific system or tool is in itself a premature narrowing of focus. This "concept = tool" approach encourages exploration or evaluation of only one or a small number of ideas. Therefore, it is unlikely that a truly different approach will be considered. This is because new approaches need some time to be refined and mastered in order to be competitive. In their immature state, it is unlikely that their performance will equal or exceed that of existing or more mature approaches or capabilities. If forced to "compete" before they are mature, it is likely that they will be discarded before they have a chance to reach their potential. As a result, DoD forgoes a level of performance that far exceeds the chosen best of breed.

To counter this tendency to go with the first approach that appears promising, DoD organizations need to undertake carefully crafted, open-minded campaigns of experimentation that offer ample opportunity in their early stages to consider a wide range of alternatives and remain flexible in implementation, taking them where the results point rather than where they were thought to be headed at their early stages.

Of course, one may attempt to avoid the choice of focus—near-term versus mid- to long-term—and create different campaigns or a basic campaign with "product spin-offs"[27] having near-, mid- and long-term implementation targets with more attention being paid to considering a wider range of alternatives (a relaxation of assumptions with regard to organization, system capabilities, and doctrine). Whether this multi-pronged approach works depends on how resources are allocated. If the

---

[27] Dubik, "Joint Concept Development." 2004.

current tendency to shortchange the activities focused on the mid- to long-term continues, these efforts will also suffer reduced probabilities of success. However, when properly managed, maintaining parallel campaigns focused on near/mid/long-term capabilities is potentially very productive.

CONFINING EXPLORATIONS TO THE INDUSTRIAL-INFORMATION AGE BORDER

Experience is both a virtue and a curse. Experienced individuals understand the tasks that an organization is currently called upon to undertake. Their understanding of these tasks and their ability to perform them well is a matter of both pride and self-identification. Of necessity, moving away from the Industrial Age–Information Age border will put individuals and organizations in an environment that becomes increasingly different, a place where a potentially different set of tasks needs to be performed or where existing tasks need to be approached differently. This brings up the question of who are the most appropriate participants in discovery activities, and who should evaluate the ideas that are generated. The history of innovation[28] suggests that the individuals who make an organization (or a profession) successful have so much invested that they find it difficult to discover or even appreciate disruptive ideas or technologies.

The transformation of DoD clearly involves an altered mission space/environment where the likelihood and the nature of the tasks to be undertaken have shifted from traditional, symmetrical combat to counterterrorism, peace enforcement, and nation building. An equally dramatic shift has taken place in

---

[28] Christensen, *The Innovator's Dilemma.* p.4.

the rules of engagement that apply and the public's expectations. Although the ability of our force to cope with these changed conditions has been admirable, working in this new mission space has proven to be challenging and uncomfortable for many.

Clearly there is new knowledge to gain, new approaches to be thought of, and new competencies to develop. These will require journeys deep inside the landscape of the Information Age. Encouraging these explorations will require that we include a variety of participants, some of whom may be unfamiliar with traditional missions and ways of doing things. Involving interagency, coalition, state, and local governments as well as private volunteer organizations, industry, and international organizations provides an opportunity to harvest multiple perspectives and learn about meaningful differences in culture and perspective so that we can find better ways to organize and conduct these kinds of operations. It will also require that the potential of new ideas and technologies be evaluated by a group more diverse than we currently employ.

TRIAL AND ERROR V. INFORMED BY THEORY

Improvements can certainly come about by trial and error, but progress will be unsure, inefficient, and relatively slow. But even a trial and error approach to improvements requires a way to measure value and associated instrumentation. This in turn requires the rudiments of "theory" and a corresponding conceptual model.

Early DoD experimentation activities relied more on trial and error and "measurement by anecdote" than on theory-based exploration and empirical measurement. While we are improving, we need to employ the rich theoretical foundation that is provided by the theory of Network Centric Warfare and related scientific disciplines that study the cognitive and social domains.

The importance of anchoring a campaign of experimentation with a conceptual model cannot be over-emphasized because a conceptual model provides suggestions as to where to look, what to look for, and how to measure or characterize what is observed.

FAILING TO CAPITALIZE ON THE CREATIVITY OF THE FORCE

We are blessed by having bright, adaptive, and creative soldiers, sailors, airmen, marines, and civilians—those at the edge of the organization who for the most part successfully grapple with new situations and problems on a daily basis. They make things work with the tools and materials they have, creating imaginative workarounds with a can-do will-do attitude. We owe it to them to remove the obstacles that limit their creativity and to capture what is being done and use it to inform our more formal experimentation activities.

CHAPTER 3

DoD TRANSFORMATION

WHY TRANSFORM?

Transformation is often driven by a need to remain competitive. This implies that change is occurring at a rate that exceeds the ability of an organization to respond while conducting business as usual.

In the Transformation Planning Guidance, Defense Secretary Donald Rumsfeld noted that transformation is, at least in part, a response to asymmetric adversaries who have profoundly changed the security landscape and who continue to evolve rapidly. He describes *transformation* as:

> a process that shapes the changing nature of military competition and cooperation through new combinations of concepts, capabilities, people, and organizations that exploit our nation's advantages and protect against our asymmetric vulnerabilities to sustain our strategic position, which helps underpin peace and stability in the world.[29]

---

[29] Office of Force Transformation. "Transformation Planning Guidance." 2003.

NATURE OF DoD TRANSFORMATION

DoD transformation is multidimensional. It involves changes in the nature of the missions that militaries are called upon to undertake, changes in the way operations are planned and conducted, and changes in the business processes that create the capabilities necessary to conduct operations. These business processes include both direct and indirect support to operations. Thus, as described by DoD's Office of Force Transformation,

> Transformation is foremost a continuing process. It does not have an end point. Transformation is meant to create or anticipate the future. Transformation is meant to deal with the coevolution of concepts, processes, organizations, and technology. Change in any one of these areas necessitates change in all. Transformation is meant to create new competitive areas and new competencies. Transformation is meant to identify, leverage, and even create new underlying principles for the way things are done. Transformation is meant to identify and leverage new sources of power. The overall objective of these changes is simply: sustained American competitive advantage in warfare.[30]

In order to accomplish a mission or task, a set of interrelated capabilities is needed. This collection of required capabilities can be thought of as a mission capability package (MCP).[31]

---

[30] Cebrowski, "What is transformation?"

[31] Alberts, *Information Age Transformation.* p. 74.
   See also: Alberts, *The Unintended Consequences.* p. 50.

To a large extent, DoD transformation is the military adaptation to the Information Age. Accordingly, the changes that are underway in DoD largely involve the application of information-related technologies that improve the quality of available information, enhance our ability to exchange information, and enable interactions and collaboration in a virtual environment. These capabilities are fundamental to developing network-centric approaches to military operations as well as the processes and organizations that support those operations and their extensions into the realms of conflict prevention and post-conflict stabilization and recovery.

Mission capability packages are the currency of transformation. One of the differences between disruptive and sustaining innovation is the number of elements of a MCP that are impacted. Sustaining innovation often involves a major change in one element of a MCP and perhaps a modest change in other elements, while disruptive innovation involves major changes in at least two elements of a MCP and more often than not changes (as a result of coevolution) in almost all of the elements of a MCP.

BALANCING THE RISKS

Change, uncertainty, and risk are inherent in transformation. Change, insofar as it involves uncertainty, is risky. The greater the change is (the greater the deviation from business as usual, the greater the departure from tradition), the greater the risk involved becomes. That is the common wisdom, but it is not necessarily correct.

Common wisdom—that incremental change is a safer and less risky approach—breaks down when the operating environment changes (1) in ways that make current practices ineffective, and/or (2) at a rate that cannot be matched by incremental adjustment processes. In the case when one or both of these conditions exist, an organization that fails to adapt is facing certain obsolescence and inevitable failure. At some point, the organization will no longer be competitive. In the case of operational military forces, this failure to adapt effectively could include a catastrophic failure of our military or of the national security infrastructure. This makes the risks associated with exploring and pursuing fundamental, disruptive changes the lesser of the two risks, and thus worth the costs involved.

The capabilities, mindsets, characteristics, and practices of Industrial Age militaries are ill-suited for current and future security challenges. Incremental improvements to Industrial Age organizations will not transform them into Information Age organizations. The changes required are simply too fundamental in too many dimensions. For example,

> the Industrial Age principles and practices of decomposition, specialization, hierarchy, optimization, and deconfliction, combined with Industrial Age command and control based on centralized planning and decentralized execution, will not permit an organization to bring all of its information (and expertise) or its assets to bear. In addition, Industrial Age organizations are not optimized for interoperability or agility. Thus, solutions based upon Industrial Age assumptions and practices will break down and fail in the Information Age. This will happen no matter how well-intentioned, hardworking, or dedicated the leadership and the force are.[32]

---

[32] Alberts and Hayes, *Power to the Edge.* p. 56.

Furthermore,

> the effectiveness of an Industrial Age organization depends upon the decisionmaking ability of one person (or a small number of persons) at the center (top) and the ability to parse and communicate decisions, in the form of guidance, to subordinates such that their actions are synchronized. Thus, centralized deliberate planning has been the traditional focus of command and control systems. Early in the Information Age, information technologies were employed to incrementally improve this traditional command and control process. With NCW, there has been a focus on replacing the traditional command model with a new one—one based upon self-synchronization enabled by shared awareness.[33]

Early returns support the promise of a network-centric approach.[34] Research sponsored by the Office of Force Transformation shows improved performance with better networked forces in Navy Fleet Operations, Army Brigade-level operations, Special Forces operations, U.S.-U.K. coalition operations, NATO peacekeeping operations, air-to-air conflict, and air-to-ground targeting.[35]

Finally, while the demands of the twenty-first century national security environment require a great deal of agility, current Department of Defense

---

[33] Alberts, *Information Age Transformation.* p. 33.

[34] University of Arizona. Decision Support for U.S. Navy's Combined Task Force 50 during Enduring Freedom.
   PA Consulting Group. Joint U.S./U.K. Combat Operations in Operation Iraqi Freedom.

[35] The Office of Force Transformation. Online Library.

organizations and strategic planning methods are too large, far too stable, and contain very limited dynamic hedging. Large, stable, analytical tools developed in the 1960s were introduced to DoD by then Secretary of Defense Robert McNamara. The highly optimized product efficiency models were based on Industrial Age metrics and timelines.[36]

Thus, significant changes are required if DoD is to meet current and future security challenges, and the risks associated with moving to an Information Age organization cannot be avoided.

But these risks can be managed. A process of coevolution that is inherently iterative and inclusive will help to expose and address the kinds of disconnects that are the root cause of the adverse consequences associated with previous insertions of information technologies. As a result, the risks associated with transformation can be reduced and the ability to recognize and take advantage of opportunities increased. This is why so much emphasis is placed upon the coevolution of MCPs.

The transformation of the DoD that is currently underway is an organizational response to the emerging threats and mission environments of this century and an adaptation to the concepts and capabilities associated with the Information Age. The theory of NCW, an expression of Information Age principles in the military domain, opens up a new territory to be "surveyed." The exploration of this landscape has begun. We have crossed the boundary between the concepts and approaches of the Industrial Age and those of the Information Age, but have, to this point, kept rather close to the border.

---

[36] Glaros, "Real Options for Defense." 2003.

The territory to be explored, the landscape of network-centric operations, is multidimensional with one set of axes corresponding to the components of mission capability packages and another set of axes corresponding to the value propositions embedded in the tenets of NCW. A point in the first set represents a particular coevolution of a MCP that in turn determines a set of points in the second set of axes, which taken together represent the fitness value for that instantiation of the MCP. Specifically, a given coevolved MCP will have associated with it a given level of information sharing, quality of information, degree of collaboration, shared awareness, and self-synchronization. Another set of axes is needed to represent the mission space. In order for a MCP to be successful across the mission space, it will need to be *agile*.[37]

Thus our explorations of this landscape (coevolved MCPs, NCW measures of value, and measures of mission effectiveness) will need to be quite varied. For example, we will need to understand the relationships between the characteristics of MCPs and the tenets of NCW.[38] We will need to understand how we can achieve desirable levels of information sharing, shared awareness, and the like. We will need to understand the relationships among the variables that form the tenets of NCW. We will need to understand how the quality of information affects shared awareness or how the nature of the collaboration that takes place improves information quality. We will also need to understand the extent to which high levels of, for example, information quality and/or shared awareness, are important for different missions under different conditions. Knowledge of general relationships, while critical to progress,

---

[37] Alberts and Hayes, *Power to the Edge*. pp. 123-164.

[38] Alberts, *Information Age Transformation*. p. 7.

will need to be augmented by specifics related to missions. For example, what characteristics, in what proportions, give us the ability to be agile over a large portion of the mission space?

This transformation will, in reality, require many different kinds of knowledge and will be comprised of many innovations. This knowledge and the innovations that this knowledge enables will allow us to translate NCW and Power to the Edge theory into practice.

Given the enormous size of the landscape to be explored, experiments will need to be conceived, conducted, and their results used in every DoD organization, large or small, military or civilian.

As we proceed on our journey of transformation, we will gain insights regarding the nature of the mission challenges we are likely to face, the characteristics of the force we need, and the processes required to create the capabilities we need. We will also gain experience, and with it, understanding of how to organize and operate.[39] Central to this will be the development of new approaches to command and control that are appropriate for Information Age missions, Information Age organizations, and Information Age environments.

INFORMING DoD TRANSFORMATION

Experimentation is part of both the art and science of transformation. The art involves the creative spark, the discovery or invention of something new. The science involves comprehending the full implications of the new ideas (as well as their

---

[39] It is likely that there is no one best way (or that it is essentially unknowable). We expect that best in this case means an approach that is both effective and agile.

limitations) and the process of systematically bringing new ideas to fruition and applying them to improve the state of the practice.

THE ROLE OF EXPERIMENTATION IN TRANSFORMATION

DoD has approached its goal of transformation with a sense of importance and urgency. As a result, JFCOM, the Combatant Commanders, the Services, and Agencies all are undertaking significant experimentation activities focused on achieving results on an ambitious (sometimes perhaps too ambitious) schedule. Despite the focus and effort already being brought to this endeavor, it is insufficient for the task at hand. We need to experiment on an ever-larger scale if we are to explore the NCW landscape adequately. We will need to engage in exploration of NCW theory at the same time that we are exploring applications of that theory. We must be building the foundations for more innovative and mature applications of NCW as we are developing and testing ways to improve the state of the practice in the near term.

CHAPTER 4

INNOVATION IN DoD TRANSFORMATION

The first step in a process of change involves recognition of a problem or an opportunity. The process then requires that the problem or opportunity is understood as the development or birth of an idea (solution, approach). In many institutions innovation fails because there is no process that serves to nurture, test, mature, and bring the idea to fruition or to institutionalize the application of the idea or the implementation of the solution as a set of changes to a MCP. A successful process of change thus requires more than experimentation. It requires an environment that supports a process of innovation.

A significant number of innovations will be necessarily disruptive. These innovations, if adopted, will profoundly affect nearly all of the individuals and organizations within DoD, changing how they think, what they do, and how they do it.

Although many who have written about transformation[40] have stressed the need to go beyond business as usual, "think outside the box," and develop new concepts of operations,

---

[40] Cebrowski, "What is transformation?"
   JFCOM. "What is Transformation."
   Alberts, *Information Age Transformation*. pp. 7-12.

a lack of understanding of what transformation really means remains fairly widespread. All too often, transformation is confused with modernization. All too often, transformation efforts are inwardly focused. Organizations claim that they can transform themselves in isolation. The focus is on how we operate rather than on how we can work with others to create opportunities for synergy. The recognition that transformation is inherently Joint and coalition has not yet reached critical mass. In the Information Age, Jointness is not an appliqué but an inherent property of everything we do. In many quarters, there is still much resistance to sharing information, to increasing the reach of collaboration, and to greater integration.[41]

In general, these misunderstandings are a result of a failure to recognize that all innovation is not the same, and that DoD transformation requires *disruptive* rather than *sustaining* innovation.

Unfortunately, innovation is currently stifled as much as it is rewarded. This needs to change. A look at the talent that leaves the military because of a perceived (and often real) lack of opportunity needs to be undertaken. Corrective measures to address this brain drain need to be expedited. Promotions based upon old core competencies do not provide the DoD with the talent it needs in the Information Age. Moreover, it discourages those with the talents the DoD needs.[42]

---

[41] Alberts, *Information Age Transformation*. p. 14.

[42] *Ibid.*, p. 14.

NATURE OF INNOVATION

Innovation, the introduction of something "new and unusual," is at the heart of transformation. Successful innovation has two parts—discovery or creation and engineering or application. [43]

> In its most general sense, innovation covers the process of bringing a new product or service into existence. The process starts with creativity blossoming in a supportive environment and ends with the launch of a successful product or service. Many recent innovation projects have concentrated on the second part of the process. This is where a creative idea is transformed into a product through rigorous and accountable procedures... This is also an area where organizations have failed in the past and the effort expended in improving the situation is justified. Developing and carrying out controlled procedures has been shown to be beneficial in many areas of business. This is also a more tangible aspect of innovation than the process of improving creativity for instance. It would however, be dangerous to ignore the creative side of innovation simply because it is more difficult to manage. [44]

Innovation is not equivalent to transformation, the result of which is "to change completely or essentially." Many new things can be introduced to improve organizations, processes, and performance without resulting in changing an organization, process, or product completely or essentially (in DoD's case, what it does or how it does it). Thus, there are degrees of

---

[43] *Webster's II New Riverside Dictionary.* 1996.

[44] Definition of "Innovation." Applied Knowledge Research Institute. 2004.

innovation. Usually innovations are grouped into two categories: *sustaining* or *disruptive*.

Sustaining innovations amount to a fine-tuning of current ways of doing things. In military terms, sustaining innovations leave existing organizations intact and do not involve changes in self-definition or core capabilities. Existing equipments or processes are improved and/or replaced with new ones that push the existing performance envelope rather than change the criteria by which success is judged. Existing processes are "optimized." Individuals and organizations get better at what they have been doing.

Disruptive innovation is completely different. When some people hear the term *disruptive innovation*, they think of Christensen's *The Innovator's Dilemma*. Christensen describes disruptive technologies as

> technologies (that) bring to a market a very different value proposition than had been available previously. [It is noted that, at first, they] underperform established products in mainstream markets. But they have other features that a few (and generally new) fringe customers value. [They] are typically cheaper, simpler, smaller, and, frequently, more convenient to use.[45]

The failure of an organization to anticipate a disruptive idea, technology, product, or service is attributed to the fact that the new product cannot compete (at least at first) with the established product on the traditional measure of value. This value proposition is so ingrained in the organization that the potential of the new product is not recognized until it is too late.

---

[45] Christensen, *The Innovator's Dilemma*. p. xv.

There is an important difference between disruptive technologies and disruptive innovation. It is the difference between evolutionary and revolutionary change. Disruptive innovation refers to revolutionary, discontinuous change, which is distinct from the incremental or evolutionary change embodied by our current technological acquisition processes.[46] In the private sector, disruptive innovation frequently takes the form of fringe products that are attractive because of their unique qualities that differentiate them from mainstream technologies. Over time, the low cost and unique attributes of the fringe product will enhance both its popularity and the resulting profits, allowing the developer to further improve the product until it can out-compete the dominant, mainstream product.[47] In this paradigm, the firm with the ability to continuously revise its products and processes to fit its environment will survive and grow, while the firm that remains fixed in its product lines and business practices will be quickly toppled by more agile competitors. In the marketplace of competing military powers, we should endeavor to become an agile, perceptive, and proactive competitor rather than a stubborn, static fixture on the battlefield.

In order to achieve DoD transformation, we must both nurture and mature these disruptive innovations, as well as dismantle the reward structures that encourage incremental, sustaining innovations.[48] Our existing doctrine obstructs this sort of disruptive progress, and many organizations and individuals have become protective of their current ways of doing business. To ease these transitions and transformations, exist-

---

[46] Thomond and Lettice, "Disruptive Innovation Explored." 2002.

[47] "Disruptive Innovation and Retail Financial Services." BAI, Innosight. 2001.

[48] Alberts, *Information Age Transformation*. p. 114.

ing doctrine must be changed to encourage and simplify the adoption of new tools and processes. Doctrine, as a tool itself, must be "fluid and helpful, not static and restrictive."[49]

In a nutshell, disruptive innovation changes the very nature of the endeavor or the enterprise. What was important before may now be irrelevant, or at least far less important. New value propositions are created. New capabilities and processes are required. In short, conventional wisdom no longer automatically applies and existing core competencies are no longer sufficient. To succeed, individuals and organizations must leave their comfort zones—a very difficult thing to do.

There are, of course, degrees of disruptiveness. In the extreme, the individuals and capabilities that were previously assets become liabilities. The processes that were honed to perfection become the greatest impediments to progress. As a result, disruptive innovations are stifled, ignored, or more often neutered or constrained to the point that they become recast as sustaining innovations. This was the case when tanks were thought of as improved horses and computers were used to automate existing processes. It was only when these new capabilities were coevolved with concepts of operations that they moved beyond having an incremental or marginal effect to having a profound effect upon the enterprise.

The odds are clearly stacked against disruptive innovation in any organization. Ironically, the more successful the organization is, the greater the odds are against adopting a disruptive innovation. Disruptive innovation cannot succeed in the context of an established organization unless a number of prerequisites exist. These include (1) recognition of the need

---

[49] Alberts, *Information Age Transformation*. p. 122.

for disruptive change, (2) commitment by those in positions of leadership, and (3) concrete affirmative steps that create the conditions and venues that spawn disruptive ideas, allowing them to be fairly and fully tested, and facilitating their adoption and institutionalization. In addition, there must be a commitment to and process for phasing out existing capabilities, organizations, processes, equipment, and doctrine.

A FOCUS ON DISRUPTIVE INNOVATION

Disruptive ideas may come from anyone or anywhere (within or outside the organization or even the domain in question). Historically, however, some of the best ideas have come from competitors or adversaries. They need to be recognized for what they are. Clayton Christensen identifies the desktop computer, Japanese off-road motorcycles, hydraulic excavators, transistors, and HMOs as several significant disruptive ideas in the business world.[50] Michael Henessey points to the German submarine, American helicopter, and suicide bombers as similarly important military innovations.[51] All of these inventions met with surprise and even opposition at the time of their introduction, yet all proved to be as successful in the field as they were disruptive to the status quo.

This occurs because many organizations have created an environment that is not conducive to disruptive innovation. Such environments reduce the chance of individuals within these organizations conceiving and putting forth ideas that challenge conventional wisdom. These dysfunctional environments are a result of major disincentives or impediments that include a

---

[50] Christensen, *The Innovator's Dilemma*. p. xv.

[51] Henessey, *Magic Bullets*. 2003.

domineering culture, short-sighted reward structures, and lack of tolerance. Organizations that punish non-conformity, originality, or adventurousness are unlikely either to give birth to or nurture disruptive ideas. "Thinking outside the box"[52] has long been recognized as a desirable trait, but such thinking has been punished as (or more) often than it has been rewarded.

The first step that can be taken by an organization committed to transformation is to make clear that it understands the difference between incremental and disruptive innovation, to commit to considering disruptive innovation, and to reward those who propose innovative ideas, regardless of whether they all pan out. This affirmation needs to be accompanied by concrete steps that include changes in reward, promotion, training, and related policies that are antithetical to innovation and investments in education and contacts with a variety of organizations. Many individuals will consider themselves adequately rewarded by simply having their ideas get a fair hearing.

TIPPING POINT[53]

Disruptive change, as it is a challenge to the status quo, is a revolution of sorts. It succeeds when and if a tipping point is reached—where the forces and momentum associated with the change amass more force than that associated with the status quo (including the forces that are proponents of "modernization").

---

[52] Phrase coined in 1995 by creative thinking guru Mike Vance in his book *Think Out of the Box*. Other titles by him include *Raise the Bar* and *Break Out of the Box*. Vance and Deacon, *Think Out of the Box*. 1995.

[53] The tipping point is the point at which the rate of change increases dramatically.
See: Gladwell, *The Tipping Point*. 2002.

Ideas with significant potential often die on the vine. Thus, the second step is to create a set of processes that can take ideas with potential and develop them as necessary to reach their tipping points.

There are many factors that influence the timing and nature of a disruptive change. First, there is the maturity of both the idea and its instantiation. Figure 3 depicts the temporal dynamics of innovation. Initially, a new concept or capability may have potential, but has not been refined and/or "productized" to the point that it is clearly more effective, reliable, and/or affordable than the capability that it will displace. At this point, there may be a number of individuals and/or organizations that are willing to adopt and improve the new concept or capability. These are called the *early adopters*. Even after many of the bugs have been worked out and the innovation is ready in terms of achieving a certain level of performance, its cost-effectiveness may depend upon the number of adopters and the value that is generated. This is certainly the case with network-related capabilities and applications. More individuals and/or organizations adopt when the benefit to cost ratio becomes greater than one, or more accurately, when it is perceived to be. The tipping point (i.e., when the rate of adoption and the corresponding increase in the benefit to cost ratio increase dramatically) occurs when a critical mass is achieved in terms of the nature and number of adopters and the value created. Late adopters usually need to be forced either by pressure from above, peer pressure, or by a discontinuation of the previous product, service, process, or approach. When the tipping point is reached, the pressures to adopt become greater than the pressures to maintain the status quo.

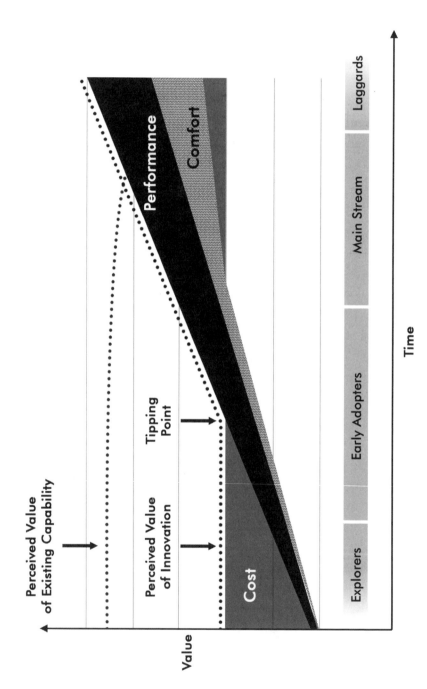

Figure 3. Temporal Dynamics of Innovation

A critical mass can be achieved in a number of ways. Experimentation is an essential ingredient in all of them.

EXPERIMENTATION AND DISRUPTIVE INNOVATION

A campaign of experimentation designed to result in disruptive innovation will need to place more emphasis on variety and develop a "story" that convinces skeptics of the soundness of the new approach. Planners of this kind of experimentation campaign will not be able to specify all or even a large fraction of the particular experiments that will be needed, nor will they be able, with any assurances, to specify a time table for moving from one phase of the campaign to the next. Moving from one phase to another will need to be determined by progress, not by a schedule. Trying to move too quickly will likely result in pushing the campaign from one that seeks disruptive change to one that settles for sustaining innovation.

Recognizing that there are considerable pressures for results, the overall experimentation program supporting DoD transformation should have activities devoted to both sustaining and disruptive innovation so that progress can be made while the search for a breakthrough goes on. Ideally the return on investment for sustaining innovation (sometimes called modernization) offsets the ongoing costs associated with seeking disruptive innovation. But even if they do not, the DoD cannot afford the risks associated with a failure to transform. Thus, the same set of expectations should not be applied to campaigns of experimentation that are focused on transformation as are applied to those that focus on modernization.

Some believe that one can focus a campaign on transformation and spin off near-term incremental improvements; however, more often than not, these competing goals are not well-managed and the efforts fail, usually by sacrificing the long-term transformational objectives.

The early stages of a transformational campaign of experimentation need to include a fairly large number of discovery-oriented activities, often involving individuals and organizations who are unconstrained by conventional wisdom. It is interesting to note that individuals who grew up with the Internet and with instant messaging have developed a set of skills that are not common among individuals who did not. DoD currently relies almost exclusively on individuals who are not of this "Internet generation" to design and evaluate new capabilities. We do this because we value their military experience. The problem is obvious. This assumes that the military of the future will be more like the military of the past than the world of today and tomorrow. The easy solution is to hedge our bets and employ a variety of people to help us generate, test, and evaluate new ideas. Clearly, those steeped in existing ways of doing things are, in general, better suited for campaigns of experiments that are focused on sustaining innovation than disruptive innovation. This is not to say that age alone determines the ability of an individual to think outside the box, but that it would be a mistake to rely solely on a homogeneous group for new ideas or input to the evaluation process.

A campaign of experimentation needs to do more than generate a variety of ideas; it also needs to select which ideas are worthy of further consideration and ultimately describe these ideas in such a way that they will be adopted, at least by "pioneers" or early adopters. In order to do this, a conceptual framework is needed that can translate the characteristics of the idea into a value proposition. As indicated earlier, the value proposition for a disruptive capability may be quite different from the value proposition for a sustaining innovation. At any rate, the value proposition for a sustaining innovation is well-established while the value proposition for a disruptive innovation may need to be developed as part of the campaign of experimentation. For example, a value proposition for a sustaining innovation will, in all likelihood, be related to mission success (for a given mission or set of missions). By contrast, the value proposition for an Information Age DoD concept might focus on agility.[54]

Variety is also important with regard to the organizations that need to participate in the experiments and analyses. Remembering that with disruptive innovation, more elements of a MCP are in play than with sustaining innovation, there is a need to have a wide range of relevant communities involved.

Thus, for a number of reasons, campaigns of experimentation that can produce disruptive innovation and hence support transformation differ in significant ways from campaigns that are focused on producing sustaining innovation.

CREATING A CLIMATE OF INNOVATION

Many successful organizations fail to be fertile breeding grounds for innovation and experimentation. Being a leader in your field or a highly effective organization is often a significant deterrent to change. Enterprises that believe they are the best in the world (or their industry) are often unable or unwilling to take the necessarily disruptive steps required to move beyond existing competencies. In the language we have

---

[54] Alberts and Hayes, *Power to the Edge*. pp. 123-128.
   Office of Force Transformation. "NCO Conceptual Framework Ver 2.0." p. 2.

adopted here, they are happy with sustaining innovation, but not at all ready for disruptive innovation. The U.S. military, widely acknowledged to be the best in the world, can be expected to experience this difficulty.

While the Department of Defense, and particularly its most visible leaders, have committed themselves to a process of transformation and have identified experimentation as an essential element of their strategy for transformation, the question of whether the conditions necessary for innovation have been created remains unanswered. This is particularly true of the disruptive innovation required to transform the Department from an Industrial Age military with Information Age capabilities to an Information Age DoD.

Indeed, one of the most common challenges that arises when the ideas of innovation and experimentation are espoused is resolving the question of what must be done to create a climate where these activities are valued and likely to be successful. The military is, quite properly, fundamentally a conservative organization (one that changes only slowly and only when there is an overwhelming reason to do so) because it is charged with the defense of the nation, including the use of lethal force, and seeks to avoid unnecessary loss of lives and waste of national resources. The issue here is, therefore, how innovation and experimentation can be most effectively fostered in such an institution. Furthermore, disruptive innovation and transformation involve creating new elites. This is also very difficult in a conservative institution. However, the cavalry had to be supplanted by armor and battleships by aircraft carriers in order to maintain U.S. national security. Neither change was easy, in part because they involved changes in the military elite.

Similar issues must be confronted to move from an Industrial Age military to an Information Age military.[55]

This section defines the issues involved in creating the conditions for innovation and experimentation and then offers thoughts on how those conditions can be created.

CLIMATE OF INNOVATION

Innovation means doing new things or doing old things in new ways. Hence, a climate that fosters innovation sets the conditions necessary to develop and explore new ideas and organize them into sets (sometimes blending them with some older ideas) such that important goals can be achieved. In some cases, those goals could not be achieved before (doing new things). In other cases, the gains are in effectiveness or efficiency (doing old things in new ways).

A climate that fosters experimentation is one in which there is no presumption that the "best" or "correct" approach is already known or can be prescribed a priori. Hence, it is a climate within which new ideas and approaches are to be developed and assessed through a combination of well-disciplined intellectual exploration and experimentation in realistic applications. This climate expects that ideas and approaches (innovations) will change as we learn about them. It also expects that some ideas will prove incorrect and will need to be either abandoned or altered in fundamental ways before they can be found worthy. However, it recognizes such errors as a

---

[55] For a detailed look into issues that must be confronted to achieve transformation, and eight key points of failure in that process, see: Kotter, "Leading Change." 2000.

natural and proper part of the effort and does not punish or denigrate those efforts.

Climates appropriate for both innovation and experimentation will be necessary for military transformation, whether in the U.S. or around the globe. New ideas will need to be put through the crucible of assessment and systematic evaluation before they are ready for implementation.

• PROCESSES OF INNOVATION AND EXPERIMENTATION

Successful innovations pass through four phases. First, ideas need to form. Second, the potential of these ideas needs to be recognized. Third, the ideas need to be formulated into innovations that can be refined, explored through experimentation, and matured. Finally, these new ideas need to be implemented and institutionalized. However, these stages should not be thought of as independent. Rather they are linked, interdependent, and often occur as iterative processes.

Ideas are the essential raw material of innovation. Some organizations behave as though there is nothing they need to do to spawn creative and potentially effective ideas. They appear to believe that ideas, like germs, are all around us all the time and cannot be avoided. While there are probably ideas being spawned spontaneously, these organizations fail to recognize their often ill-formed expressions as such and make no serious effort to properly articulate them or to differentiate valuable and important ideas (potential for disruptive innovation) from others (sustaining innovation or unimportant ideas). Other organizations realize that ideas are like ore—of differing quality and importance. They see important ideas as needing to be stimulated, cultured like plant seedlings, and valued. They will both create venues where ideas can be surfaced and encourage their production. Ultimately, these organizations will develop a strategic competitive advantage.

How ideas are selected for consideration is also important. Mechanisms, both routine and ad hoc, for surfacing new ideas and offering them to the interested community for thought and reflection, are critical in this process. Establishing routine mechanisms to accomplish this is a way of signaling the organization's valuing of and interest in identifying promising ideas. These may also include ongoing "competitive intelligence" efforts to understand what new ideas are arising in the field and what adversaries are doing and thinking. Ad hoc processes, which might range from senior and middle managers asking for new ideas when they are interacting with individual employees to temporary committees or task forces created to deal with perceived or emerging challenges, can also be important because they signal a continuing interest in constructive innovation. Open door policies fall in this same category, provided that they are genuine.

Taking ideas and organizing them for thoughtful assessment and refinement is the third essential step. This step must be resourced intelligently. Ideas being translated into concepts and capabilities will need protection as well as rigorous experimentation and evaluation. The original formulation will almost never be directly usable in the operating environment. This is the stage where campaigns of experimentation, including the agility they require for success, will be most important. This is also the stage where decisions must be made about the largest investments. Those investments may actually occur in the last stage (implementation and institutionalization), but they are determined here. This is also the stage where coevolution is most important. Innovation in a single arena, whether that is doctrine, organization, technology, training, leadership, or any other isolated area, will almost certainly fail and will always fall far short of its potential if it is not coevolved with complementary change and innovation in other areas.

The transition to and effective execution of the last stage is, however, perhaps the largest barrier for ideas that create disruptive innovation. Within large organizations (and DoD is one of the largest in the world), bringing disruptive innovation ideas from the proof of concept and prototyping stages into general use is often the most significant challenge. The classic model of innovation moving from explorers to pioneers (early adopters) to reach a tipping point where broad adoption occurs is much easier to follow in an open market with a variety of actors, each of whom has relative freedom, than in a heavily segmented organization with multiple decision (and potential veto) points and potentially drawn-out processes. This is, of course, the place where leadership is most crucial, particularly when we are talking about a Joint force that must work together coherently during any transition period and where key parts of institutionalization are, by law, still heavily decentralized and, by practice, implemented in very different operating environments around the world.

• RECOMMENDATIONS

The key practical recommendation for fostering a climate of innovation and experimentation was summarized by a leading practitioner at an IAMWG session as "Leadership, leadership, leadership!"[56] Leadership for this purpose can be summarized

---

[56] Donald G. Owen, a retired U.S. Navy Captain with a long record of successful innovation as a C2 analyst. December 14, 2004.

as envisioning the future and communicating each individual's role in it. Leaders who set the tone that innovation is not simply OK, but rather expected and valued, make a huge difference in the likelihood that new ideas of value will surface, be recognized, developed, and implemented. The recent OFT case studies on Network Centric Operations found leadership central in the successes of Task Force 50 in the Navy, NATO peacekeeping operations in Macedonia, Special Warfare Group 1 in Afghanistan and Iraq, and the Stryker Brigade.[57] Leaders must be able to go beyond "talking the talk" to "walking the walk" in order to accomplish successful disruptive innovation within the military.

A second key factor is the breadth of participation within an organization. While standardization is an effective tool for helping people generate shared understanding in a familiar environment, diversity of perspectives and experiences is invaluable in ensuring that conventional wisdom is challenged and that new opportunities are recognized. The needed diversity here goes beyond the legally mandated issues of race and gender. It involves people from multiple generations, with multi-disciplinary educations, with different training, and with various concepts of success.

This means creating and supporting rich communities of interest. This occurs within a service like the U.S. Army when combined arms teams are created so that the opportunities for

---

[57] University of Arizona. Decision Support for U.S. Navy's Combined Task Force 50 during Enduring Freedom.
Reinforce. Multinational Operations (During IRTF (L) trial of AMF (L); Amber Fox; and ISAF 3).
PA Consulting Group. Joint U.S./U.K. Combat Operations in Operation Iraqi Freedom.
RAND. Stryker Brigade Combat Team.

the use of different types of assets are understood within the core team. The need for jointness in thinking, force development, operational planning, and execution is further recognition of the need for different perspectives. Even jointness, as it is currently envisioned and practiced, does not begin to utilize all of our information and all of our assets. It is clearly inadequate in an operating environment characterized by coalition operations and the need to support non-military organizations and institutions in missions both at home and abroad. As the U.S. focuses on how to avoid conflicts (whether by deterrence or diplomacy), win wars when they must be fought, and win the peace, it becomes important to recognize the importance of involving all the members of the Joint force, coalition partners, the interagency process, industry, and non-governmental organizations in order to enrich the set of perspectives available for innovation and experimentation. Even those organizations where innovation and experimentation are an integral part of their mission (DARPA, battle laboratories, etc.) are largely staffed by people from one discipline—usually engineering. Social scientists, and even operations research specialists, remain in short supply when the needs for coevolution across the lines of development (doctrine, organization, training, etc.) are considered. Broadening the set of available perspectives is essential if groupthink is to be avoided and innovation encouraged.[58]

Communities of interest are needed so that theoretical and the practical perspectives can be adequately considered. Good, abstract ideas are often the beginning point for good, effective practices. Analysis after analysis shows that new ideas are improved by exposure to those who will have to use them as

---

[58] Janis, *Groupthink*. 1982.

ideas are transformed into capabilities. Hence, it is important that leaders foster the construction of broad-based teams to create a climate for innovation and experimentation.

Beyond ensuring that a variety of perspectives are created, good leadership will also ensure that individuals with new ideas and dissenting perspectives are valued and recognized. All too often the "reward" for advocating new ideas within the Department of Defense is early retirement. Better leadership would specifically recognize those who challenge existing wisdom and ensure that they are given assignments that challenge them to demonstrate the value of their ideas and the means to test them.

Currently there is little or no recognition of the value of those civilians and serving officers who develop the skills necessary to plan, organize, and conduct effective experiments. Not surprisingly, few individuals with those skills can be found within the Department of Defense. Linkages to those institutions where these skills are located (for example, the Army Research Institute, Office of Naval Research, Air Force Human Resources Laboratory, and the Army Research Laboratory) are often weak.

Innovation and experimentation can also be fostered by a variety of organizational steps. Within industry, particularly rapidly changing sectors such as information technology, the process of innovation is often encouraged by creating small units whose sole purpose is to look for new ideas that can or will destroy the existing approaches. The famous Lockheed "skunk works" was one such effort within the defense industry, though it was created as much or more for security reasons as it was to foster innovation. The Project Alpha effort in JFCOM is a similar effort, though apparently intended as much or more to ensure that important new ideas are recognized as to generate them. Large size appears to be a detriment to innovation, so the creation of small, focused, and specialized entities with charters for innovation should be the norm. Beyond creating small groups and charging them with achieving innovation, Jack Welch, famous for his leadership of General Electric, also called for "boundaryless" organizations, encouraging those groups to reach out to anyone else in the company who might help them.[59]

Similarly, businesses often create small groups focused on competitive intelligence or examining what competitors (both large and small) are bringing to the marketplace in order to ensure that they avoid surprises in the marketplace. However, within the Department of Defense, these roles are assigned to the intelligence community and do not closely interface with those charged with developing the future force—whether that is considered the Combatant Commanders, the Joint Staff, the Services, or JFCOM. Intelligence remains largely "stovepiped," reporting up its own chains of command. The process of integrating intelligence as it moves up in the organization also tends to squeeze out dissenting or minority opinions; yet disruptive change in the competition is most likely to be recognized by the "dissenters"—particularly early on. Lack of horizontal linkage has made it difficult to "connect the dots" when new developments emerge. While intelligence reform is now underway, those efforts do not appear to focus on recognition of military innovations, changes in the threat around the globe, or the need for new capabilities within U.S. forces.

However, the single most important challenge facing leadership that wants to foster innovation and experimentation is the lack of flexibility in resource allocation. The existing processes

---

[59] Slater, *Jack Welch & The G.E. Way.* 1998.

are designed around Industrial Age standards and unrealistic expectations about the predictability of campaigns of transformational experimentation. When a new idea emerges, few resources are available to "massage" it into a workable form. Even when this occurs, it may be years before funds and venues to explore it are available. When a campaign is launched, the resources to rapidly exploit early successes are seldom available. Resources tend to be consumed in large "venues" for experimentation—exercises that are sometimes linked to force training goals (which limits the degree of innovation that can be introduced) and that always consume major resources, leaving relatively little for the smaller, more focused efforts (for example Limited Objective Experiments, Modeling and Simulation Experiments, and Laboratory Experiments) where greater gains in knowledge are often possible. Resource constraints have tempted too many to attempt to piggyback experiments on training exercises. This invariably compromises the experiment.[60]

Finally, quality of experimentation has often been limited when leadership is willing to settle for relatively low rigor in general and particularly in the metrics being used and the controls in place. The cost of properly instrumenting an experiment, particularly a human-in-the-loop experiment, can be very high. As a result, there is a tendency to rely on surveys that focus on the satisfactions and perceptions of the participants and subject matter experts (whose expertise is typically based on historical experience, not the innovations under study) rather than measuring behaviors and comparisons with formal baselines. These practices not only make it less likely that disruptive innovation will be rated as successful, but also

---

[60] Alberts et al., *Code of Best Practice for Experimentation.* p. 56.

send a clear message to the participants that innovation and experimentation are not genuinely valued and important.

In summary, then, leadership is the crucial ingredient in creating a climate for innovation and experimentation. The issues leadership must be willing and able to address include:

- The paradox of "groupthink" cells versus the proper balance of expertise. Being careful not to create a brain drain due to a lack of funding for personnel;

- Elaborating more on the value of long-term experimentation;

- Ensuring breadth of participation within the organization by crafting communities of interest;

- Mandating diversity of participants in the processes;

- Creating teams that balance theoretical and practical perspectives;

- Valuing and recognizing individuals for innovation and experimentation;

- Creating relatively small groups with specific charters for focused innovation;

- Building competitive intelligence organizations and linking them to the operational communities;

- Enabling flexibility in resource allocation and reallocation; and

- Ensuring the use of rigorous measurement and baseline processes in experimentation.

CHAPTER 5

CAMPAIGNS OF EXPERIMENTATION

Campaigns of experimentation are a proactive response, a way of focusing attention and resources to make something happen. The Manhattan Project[61] and the War on Cancer[62] are examples of attempts to orchestrate focused campaigns of experimentation. A campaign of experimentation is defined in the *Code of Best Practice for Experimentation* as a "set of related activities that explore and mature knowledge about a concept of interest."[63] A campaign of experimentation seeks to accomplish one or more of the following: focus attention on specific outcomes; accelerate progress toward one or more objectives; reduce risk; and/or make some process more efficient. As with "better, faster, cheaper," it is difficult if not impossible to achieve all or even a major subset of these objectives simultaneously with initial instantiations of a disruptive capability or approach.

---

[61] The Manhattan Project Heritage Preservation Association.

[62] The War on Cancer is a broad-based effort to cure the various forms of cancer. For more information, see: Waldholz, *Curing Cancer.* 1997.

[63] Alberts et al., *Code of Best Practice for Experimentation.* p. 25.

Campaigns provide a continuing opportunity to organize and extend knowledge on a given subject. Without this opportunity, the fruits of individual experiments will not be fully harvested. The *COBP for Experimentation*, focusing as it did on the conduct of individual experiments, introduced the idea of a campaign of experiments because the authors felt the need to remind readers that experiments do not exist in isolation—that they are, in fact, inherently part of a larger process of experimentation; that a single experiment can accomplish only so much (i.e., it can explore a limited number of variable interactions); and that the contribution of any single experiment is very limited unless its findings are replicated and extended in a systematic way.

NATURE OF A CAMPAIGN

A campaign of experimentation involves balancing variety and replication. In a naturally occurring process of experimentation, the amount of variety and replication is an emergent property. Because time is not controllable in the process of knowledge maturation, breakthroughs and filling in the details occur at a pace that is determined by the collective decisions of many individuals and organizations. For the most part, these people work independently. A campaign seeks, among other things, to compress time and make more efficient use of resources. As a result, within a campaign, decisions need to be made that limit variety and replication.

Although at first glance, replication may seem to be unnecessary and thus wasteful, it is an essential element in creating and maturing knowledge. Scientists value replication because they understand that "perfect" experimentation on any meaningful topic is all but impossible. Pure machine experiments, the most heavily structured and controlled types of experiments, are assemblages of assumptions that depend on a very large number of algorithms and rules. They are often run thousands of times to establish relationships. Even so, they typically produce results that are subject to multiple interpretations. Experiments that involve humans and organizations[64] (human-in-the-loop) present significant challenges in any number of arenas, including but not limited to measurement, subject selection, training of subjects and observers, control for outside factors, cultural factors (organizational as well as national), and the interpretation of results. Replication, particularly when it is conducted by different researchers with different subjects, is an insurance policy against all these types of "problems" or sources of "imperfect" (or wrong) knowledge.

In its simplest form, replication can be accomplished by ensuring that the same data are collected across a set of experiments. That is, while new experiments within the campaign may look deeper at the same concept or introduce new factors, they should also continue to collect the same variables that were included in the earlier, simpler experiments. This provides a continuing opportunity to replicate the initial findings or to discover issues for further research. For example, if an early strong finding is weaker during replication, then it is reasonable to suspect that some specific factor may be found to have been weakly controlled in one or the other of the efforts or some different condition (different types of subjects, different experimentation context, etc.) may be found to be more important than originally thought.

---

[64] DoD transformation is, at its heart, about changing individual and organizational behavior.

For machine experiments, replication is relatively simple and cost effective. For small scale, human-in-the-loop experiments, the design and drivers (training materials, scenarios, measurement tools, etc.) can be made available to researchers with limited resources (for example, professors in military schools, researchers in friendly countries) so that they can use the experiments for teaching and research purposes. The chances that some imperfection exists—whether an unrecognized factor that was not controlled, a bias built into the research design, some measurement error, or simply some random variable not likely to occur again—are simply too great to ignore.

Merely considering a "complete" set of variables is not sufficient. There is also the problem of sorting out the interactive effects among many of the phenomena of interest. This presents a methodological challenge that requires multiple experiments that collect a common set of data so that the contributions of each individual factor can be isolated. Hence, it is often wise to design a series of experiments that begins with establishing the role and strength of each factor expected to be important, then moves on to other experiments that bring them together in increasingly larger combinations so that their independent and interactive effects are highlighted. Hence, the degree of complication (number of interacting factors) of the experiments will grow over time.

While replication seeks to collect sufficient information upon which to base theoretical conclusions, multiple experiments are also needed to ensure that the applications that are developed are robust—that they apply across a range of situations. The amount of variety considered will determine how robust the findings will be. Individual experiments are, by definition, focused on a narrow set of issues and situations. This is necessarily true if they are going to be properly crafted experiments that limit their focus sufficiently to generate well-understood results. While a single experiment can be designed to include some very different situations (for example, a pre-conflict phase, outbreak of hostilities, serious combat, termination of hostilities, and stabilization operations), the size and scope of these efforts makes them relatively unusual. Moreover, that approach deals with only one dimension across which situations are different. For example, they will deal with only one set of coalition partners (or none) out of the myriad possibilities. Similarly, they will explore, at best, deploying and sustaining a very limited set of U.S. forces or the specific types of information challenges that must be faced in real-world contexts. Clearly a campaign of experimentation that includes experiments that embrace variety (e.g., situations, approaches, and participants), one that is designed to sample the interesting issue space, is the best way to ensure the agility of the approaches examined or the robustness of its findings.[65]

Another issue related to robustness is the ability to define the limiting conditions that apply to any specific set of experimentation findings. The Latin *ceribus paribus* and the English "all other things being equal" are reminders that an individual experiment is necessarily confined to a single set of circumstances. Those circumstances may involve the reliability of the equipment being employed, the level of training of its operators, the level of training of the unit for the type of mission and situation involved, the "hardness"[66] of the military teams involved, the nature of the adversary, the relevant order of battle, the weather, the length of the experimental trials, or any of

---

[65] Alberts et al., *Code of Best Practice for Experimentation.* pp. 205-207.

[66] Hardness: experience at working together on similar problems.

the other myriad factors that have been shown to make a difference. What appears to be a strong effect in one set of circumstances cannot be assumed to apply to all circumstances. Hence, conducting a carefully constructed set of experiments is necessary in order to test for the presence and impact of limiting conditions.

As mentioned earlier, when dealing with complex phenomena, research findings themselves are often subject to differing interpretations. This is particularly true when new concepts are being explored and when teams of peer reviewers with multi-disciplinary backgrounds are employed (which best practice requires). In many cases, the only way to resolve issues of interpretation is to conduct related experiments designed to differentiate between the alternatives to understand the findings. One consequence of this factor, developed in more detail later, is that campaigns of experimentation should not be fully defined in advance, but rather should be crafted to permit the introduction of new factors and new measurement approaches that prove important in the early experiments.

Finally, not all experiments produce their expected findings. Many, particularly discovery experiments and those focused on preliminary hypotheses, are likely to generate unexpected findings. These findings are important and interesting precisely because they differ from expectations. They can indicate a flawed experiment, an encounter with unanticipated limiting conditions, a measurement problem, or new knowledge in the form of unforeseen variables or interactions among variables. Campaigns of experimentation provide the opportunity to explore unexpected patterns and mature relevant knowledge.

ANATOMY OF A CAMPAIGN

Successful campaigns of experimentation must be organized around a specific focus and set of objectives. A conceptual model, specifying the relevant variables and their (hypothesized) relationships, provides a unifying framework that puts what we know and what we do not yet know in a suitable context. Associated with a conceptual model is a set of metrics that specifies the attributes for the variables of interest. Experiments that generate data, analyses that explain what the data mean, and the inferences that can be drawn from the available data form the core activities and products of a campaign of experimentation.

FOCUS AND OBJECTIVES

Campaigns of experimentation are all managed and directed to some degree.[67] Thus, at the heart of any campaign is its focus. The focus of a campaign may be general or it may be specific. Associated with a campaign's focus are its objectives— how you tell whether or not the campaign has been successful. The state of knowledge or the state of practice is almost always a point of departure for a campaign of experimentation. Campaigns may be focused on theory or its application.[68] Those focused on theory could have one or more of the following objectives: validate basic relationships, establish boundary conditions, seek other relevant variables, or establish values for parameters specified by the theory. Those focused on applica-

---

[67] This does not imply that there is a single entity in charge of a campaign. A campaign may be like a coalition operation undertaken by a coalition of the willing acting on a set of shared goals.

[68] We believe that a single campaign should not try to do both; however, related campaigns that seek to advance theory and practice in tandem can work well.

tions could have one or more of the following objectives: conceive of an application, build a prototype, or evaluate the utility of the application under a set of circumstances.

Campaigns, like experiments, can only accomplish so much. Campaigns that start with a relatively untested theory cannot be expected to test, refine, and mature that theory while simultaneously developing mature and tested applications. Such an effort would be rather difficult if not impossible to manage. Hence, experimentally informed transformations of large institutions will require multiple campaigns that will need to be orchestrated to some extent. How this applies to the transformation of the DoD is examined in the next chapter.

The objectives of a campaign of experimentation need to be carefully considered because they determine the issues or questions that will be investigated. For any given question or issue there is something that is known—a point of departure. In most cases there is also some idea or theory (or perhaps alternative theories) that offers an explanation—an idea of the relevant causes and effects, as well as key assumptions and limiting conditions. Existing knowledge about the question at hand provides the starting point while proposed theory provides an initial direction for the campaign. Together they determine the nature of the evidence that will be required to move ahead.

CONCEPTUAL MODEL

A theory, an idea, or an approach needs to be articulated in a way that facilitates assessment, testing, and exploration. A conceptual model, one that identifies the variables involved and their relationships, provides a way to rigorously explain what one has in mind, as well as a way to organize existing knowledge and relate it to that which is not known or established.

A conceptual model is analogous to a corporate balance sheet. It gives a picture of the state of our understanding and points us in the right direction. Like a balance sheet, a conceptual model will change over time, hopefully improving as the campaign unfolds. An important part of this process of maturation is the increased ability to express and differentiate the meanings of variables (or establish meaningful groups of variables). Improvements in our ability to measure exactly what we want to measure and to do it with greater precision and reliability are also an important part of this maturation process.

A conceptual model is implicit in any experimentation activity. However, making it explicit is necessary to communicate clearly what the campaign is all about, to understand better what needs to be done, to determine the order in which it needs to be done, and to measure progress.

The state of the conceptual model at any given point in time should serve to inform those who are managing a campaign. A conceptual model also provides a way to be specific about the objectives of the campaign and its achievements.

Although for many campaigns, success will be measured in terms of a tangible product, we should not lose sight of the fact that successful campaigns will also result in an improved, or more mature, conceptual model, an intangible product that may have more far reaching implications than the specific products produced. Improvements in the conceptual model will move it (1) from being vague to precise, (2) from just isolated elements of knowledge to a richly connected explanatory and predictive theory, and (3) from an expression of knowledge that is understandable only by specialized researchers to an articulation of knowledge that is understandable and appreciated by a wider audience.

EXPERIMENTS

The formulation of a campaign of experimentation is centered around the evidence that is required to develop and/or test a theory or assess an application—evidence that can provide answers to the question(s) at hand. The word *evidence* is defined as "something that furnishes (or tends to furnish) proof."[69] Thus evidence is not equated with proof, which is a logical product of analysis (a conclusion), but with the inputs to an analytical or thought process. Both observation and testimony can constitute evidence; however, not all evidence is equally relevant, valid, replicable, or credible. Properly designed and conducted experiments greatly increase the likelihood that the data collected, the observations made, or the testimony (expert opinion) elicited will have these desirable properties. Multiple experiments and analyses are required to establish relevance, validity, repeatability, and ultimately credibility.

Thus, the conduct of properly designed and sequenced experiments is integral to any campaign. These experiments will differ by the maturity of their knowledge contribution, the fidelity of the experimentation setting, and the complexity of the experiment. They will also differ with regard to the time, resources, and expertise required to undertake them. These issues are discussed in the following chapter.

---

[69] *Webster's Third New International Dictionary.* 2002.

The nature of the experiments that are appropriate for a particular campaign and at any given stage of a campaign will vary depending upon a campaign's focus and objectives relative to the state of our knowledge. The *COBP for Experimentation* defined three purposes for experiments: discovery, hypothesis testing (confirmatory), and demonstration. Each of these relates to one or more of the characteristics of the verb *to experiment* as it is commonly used. Although the nature of the evidence that each of these types of experiments can collect will vary, these different types of experiments serve to complement and build upon one another in the conduct of a campaign of experimentation. Hence, they each contribute in their own way to creating, refining, and disseminating knowledge.

DISCOVERY EXPERIMENTS[70]

Discovery experiments are designed to generate new ideas or ways of doing things. They seek to create opportunities for individuals and organizations to "think outside the box" and thus to stimulate creativity. They often involve providing novel systems, concepts, organizational structures, technologies, or other elements in a "hands on" setting where individuals and organizations can explore their use and where these "explorations" can be observed and catalogued. Discovery experiments provide an opportunity to develop promising alternatives to current approaches and systems and to refine them to the point where their potential can be assessed realistically. It is important that a new idea, approach, or system be adequately refined before it is compared to current practices or doctrine. If it is not, then the experiment will be focused on testing an immature, incomplete application (instantiation). In this case,

---

[70] Alberts et al., *Code of Best Practice of Experimentation.* p. 19.

generalizations beyond the instantiation to theory or generic applications will not be valid.

The product of a discovery experiment is a promising idea or approach. The process of discovery is somewhat familiar to all of us. We have all approached a problem by trying different solutions or approaches to see if one of them works, or which one works best. While this is the basic idea behind discovery experiments, it is important that these experiments be designed and conducted in a way that (1) encourages full exploration of the possible solution space and (2) is instrumented in such a way that permits accurate descriptions of the approaches or solutions tried and the results. Although discovery experiments do not necessarily involve the formal control of a set of variables (to isolate influences and effects) they need to provide enough data so that the "promising" approach can be compared to the status quo or some other alternative to establish potential value (or validity) and so that the approach can be replicated.

Properly performed discovery experiments help to ensure that the campaign considers a full range of alternatives and does not prematurely narrow the alternatives (e.g., approaches, explanations).

HYPOTHESIS TESTING EXPERIMENTS[71]

Hypothesis testing experiments are the classic type used by scholars and researchers to advance knowledge by seeking to falsify specific hypotheses (specific if—then statements) or discover their limiting conditions. They are also used to test

---

[71] Alberts et al., *Code of Best Practice of Experimentation.* p. 22.

whole theories (systems of consistent, related hypotheses that attempt to explain some domain of knowledge) or observable hypotheses derived from such theories. In a scientific sense, hypothesis testing experiments build knowledge.

It is generally accepted that multiple experiments of this type are needed to develop quality data in sufficient quantities in order to provide a foundation for confidently establishing new knowledge. Depending on the nature of the hypotheses tested, this type of experiment provides "proof" that a theory, idea, or approach is valid; establishes its value under specific conditions; establishes the exceptions and limits of its application or utility; and establishes a degree of credibility.

DEMONSTRATION EXPERIMENTS[72]

Demonstration experiments create a venue in which known truth is recreated. These are like the experiments conducted in a high school in which students follow instructions to prove to themselves that the laws of chemistry and physics operate as the underlying theories predict. Technology demonstrations fall into this category. They are used to show potential customers that some innovation can, under carefully orchestrated conditions, improve efficiency, effectiveness, or speed. In successful demonstrations, all of the technologies employed are well-established and the setting (scenario, participants, etc.) is orchestrated to show that these technologies can be employed effectively under the specified conditions. Immature technology or inappropriate settings or scenarios will fail to achieve the desired result. Thus, demonstration experiments are designed to convince, educate, and (at times) train.

---

[72] Alberts et al., *Code of Best Practice of Experimentation.* p. 23.

ANALYSIS

Analysis takes the data provided by experiments, combines it with previously collected data,[73] and develops findings that serve as the basis for drawing conclusions related to the issues or questions at hand. Statistical theory forms the scientific basis for determining the probability that the observed data have a given property (e.g., two treatments are significantly different) with a given level of confidence, or in other words, that there is little likelihood that the result occurred by chance. Increasingly, this analysis extends into areas of complexity[74] where analysis is more challenging and requires new approaches and tools intended to identify emergent behaviors and system properties.

Analysis needs to take place before, during, and after the conduct of each experiment. The conceptual model provides a framework and point of departure. There are many analytical techniques that can be brought to bear and care must be taken to employ the appropriate method or tool.[75] The findings developed in each of the analyses that are conducted should be used to update the conceptual model and should be disseminated to others engaged in the campaign or related campaigns.

---

[73] The ability to use data collected by others is essential to advance at a reasonable rate. This is why documentation, including metadata tagging, is such a vital responsibility of researchers and research (experimentation) organizations.

[74] Moffat, *Complexity Theory.* 2003.

[75] Each statistical method or tool is based on assumptions about the way the sample is obtained and the nature of the underlying distributions of the key variables.

The *COBP for C2 Assessment*[76] provides a prescriptive view of how analyses should be conducted in a domain central to DoD transformation to make the most of the available data and to generate additional information. The nature of the analyses required in campaigns of experimentation that are designed to inform transformation is discussed in the remainder of this book.

STAGES OF A CAMPAIGN

Campaigns of experimentation should generally move along an axis that takes them from discovery experiments to preliminary hypotheses experiments, to refined hypotheses experiments, and finally, when the state of knowledge is mature enough to support serious policy and acquisition decisions, to demonstration experiments. These purposes for experiments, defined above, are discussed in some detail in the *Code of Best Practice for Experimentation.*[77]

Another way of envisioning the path that a campaign of experimentation usually takes is in terms of exploring a knowledge landscape. In the early stages of a campaign, one is flying above the landscape where the location of each point on the terrain is defined by a vector whose elements are the independent variables and the terrain features (height) are defined by a vector of performance and/or effectiveness measures (a "value view"), seeking to find interesting parts to explore in greater depth. The amount of time and attention that is devoted to this initial exploration determines how much territory can be

---

[76] *NATO Code of Best Practice for C2 Assessment.* 2002.

[77] Alberts et al., *Code of Best Practice for Experimentation.* pp. 19-24.

covered and therefore how far from the border of existing knowledge and/or practice one can consider.

Discovery experiments are a good way to harness the imagination and creativity of individuals. How they are conceived and designed will determine the starting point for their flight of discovery, influence their initial direction, and place limits on the breadth of issues and approaches that they can explore. Discovery experiments are meant to provide the inspirational spark that gives life to a new piece of knowledge or a disruptive innovation—a spark that would otherwise not occur or occur at some unknown time in the future.

The points of departure for discovery experiments are propositions that link broadly defined concepts together. The term *propositions* is used because discovery experiments seldom begin with adequately precise definitions to formulate truly testable or refutable hypotheses. Rather, points of departure for discovery experiments involve some new idea or problem that is seen as potentially important.

The sources of discovery experiment topics may be anywhere and everywhere. They may come from individuals who are engaged in current operations, or from theorists and researchers focused on the future. Truly disruptive ideas do not usually come from those focused on incremental change. The search for ideas needs to be inclusive rather than exclusive. The search needs to include analogies based upon knowledge or research from other fields, as well as experience in other domains. Often those not yet indoctrinated in current wisdom or practices are in the best position to think outside the box. Hence, the selection of individuals will ultimately determine how far from the border a discovery experiment will venture.

In the language of the scientific method, discovery experiments are focused on observing, defining, and classifying what is occurring. They involve a great deal of creative thinking because the original idea must be translated into meaningful variables, patterns of relationships, and potential limiting conditions. In many cases, discovery experiments are also needed to develop, refine, and validate the relevant measures and measurement techniques used later in the campaign of experimentation. Data collection and data collection plans for discovery experiments must be crafted flexibly to ensure that unexpected findings are not missed and are documented.

For genuinely new topics, several discovery experiments will be needed precisely because the state of knowledge is immature, initial insights are likely to be rich, and novel factors and relationships can be expected. Moreover, these experiments are the most difficult to baseline because typically the phenomenon of interest is not precisely defined before the experiments begin. However, some existing knowledge can almost always be identified if a diverse community is consulted as the effort is conceptualized.

FORMULATING THE CAMPAIGN

The initial stage of a campaign of experimentation is focused on the identification of the territory to be explored. This encompasses the locations on the landscape and the set of value-related variables that will be considered. This is analogous to the problem formulation stage of an analysis in which the variables of interest are identified, the relationships of interest selected, constraints determined, and assumptions explicitly expressed. Some types of experimentation campaigns will involve a set of "controllable" variables. The purpose of these campaigns will be to determine the values for these variables that will result in desirable outcomes.[78] The existing body of knowledge and discovery experiments provides the data while exploratory analyses help shape the decisions involved in this phase of the campaign. It is particularly important that adequate time and resources are devoted to this phase because errors in formulation have significant downstream effects and are very costly to correct.

INVESTIGATION

The next stage of the campaign involves systematic investigation. The major activity of this phase is the conduct of experiments that test hypotheses. These experiments collect data and perform analyses that contribute to our understanding. Normally, these experiments proceed from preliminary experiments to refined experiments.

Preliminary hypothesis testing experiments build on the results from discovery experiments, existing knowledge, and established practice. They require the articulation of a clear set of interrelated hypotheses and measures that are reliable, valid, and credible. They also require the identification of a precise set of limiting conditions, which if not met therefore require more research, analysis, and discovery experiments. The task of these preliminary hypothesis testing experiments is to collect and assemble clear evidence regarding the cause-and-effect relationships at work—evidence that does not currently exist. In the language of science, the purpose is to develop explanatory or causal knowledge—to go beyond the existing idea of *what* is happening to include knowledge of

---

[78] Some campaigns of experiments may need to be conducted just to determine which measures of value have validity and reliability.

why it is happening. This is a crucial step in transformational experimentation because it provides the basis for understanding the implications of specific policies, practices, and capabilities. In most cases, this level of knowledge is needed to provide the basis for the coevolution of doctrine (including tactics, techniques, and procedures), personnel, training, equipment, organization, and the other elements necessary to field successful mission capability packages.

Refined hypothesis testing experiments build on explanatory knowledge to create *predictive* knowledge. Almost of necessity these experiments employ better definitions, measurement tools, and are more detailed (involve additional variables) than their predecessors. These experiments are the last step in the process of building knowledge (remember that demonstration experiments are predicated on displaying, not creating, knowledge). These experiments typically focus on defining the limiting conditions for cause and effect propositions developed earlier in the campaign of experimentation. In other words, they make it clear when the findings can be considered reliable and when they are likely to break down. In the language of operations research, these experiments are involved in a process of systematically relaxing assumptions to find the limits of some theoretical knowledge.

This class of experiments is also used to expose new knowledge to peer reviewers experienced and capable enough to suggest where the findings might break down. These experiments are a crucial part of a campaign of experimentation aimed at transformation because they are the last line of defense against falsely generalizing results, and thus help prevent one from moving ahead with insufficient understanding.

EDUCATION

The final stage of a campaign of experimentation moves from the acquisition of knowledge to the dissemination of knowledge. Demonstration experiments provide a useful tool for this purpose. Like the chemistry experiments conducted in high school laboratories, the goal of demonstration experiments is to show that a set of underlying propositions have predictive power and that relying on them has known consequences.

The most common DoD demonstration experiments are funded as Advanced Technology Demonstrations (ATDs) and Advanced Concept Technology Demonstrations (ACTDs). Both of these arenas explicitly call for the application of *mature* technologies in order to educate selected parts of DoD about their potential and implications. They also typically leave behind some technologies. However, many ACTDs experience difficulties because (1) the technologies are not fully mature and (2) they do not involve approaches that have been coevolved to include concepts of operations, logistical support, organizational adaptations, effective training, or other elements of successful mission capability packages. Indeed, premature ACTDs can do harm by making new approaches look weaker than their true potential. The idea of capability-based planning may prove helpful here over time by focusing attention on the larger issues being addressed rather than on a particular instantiation.

CAVEATS

There are few "pure" experiments; however, two important caveats apply to this classification of experiments (discovery, hypothesis testing, demonstration) as they have been conducted to date in DoD. First, experiments that have been conducted by DoD are not purely one of the three types. They are, in reality, hybrids that focus on exploring or communicating knowledge at more than one level of maturity. For example, it is not uncommon to see experiments that include both preliminary hypotheses for some variables and relationships while engaging in discovery work on others. Even demonstration experiments, which have a primary purpose of displaying well-understood knowledge, have been used to examine that knowledge in relatively new applications and therefore involve some important hypothesis testing. Second, not all campaigns of experimentation need to start with discovery experiments. Campaigns should start where the existing knowledge places them. For many transformational topics, there is considerable existing knowledge that makes it possible to begin work at one of the hypothesis testing levels. For example, the broad use of computer networks in business inventory management provides rich hypotheses and well-established knowledge that may apply to military logistics and management policies and practices within DoD. However, because this knowledge was not developed in a context like the Department of Defense, well-crafted refined hypothesis experiments are the most logical starting point.

FIDELITY AND CONTROL

Throughout the course of a campaign of experimentation, a variety of experimental activities will be undertaken. These differ primarily with respect to two characteristics: fidelity and control. Campaigns of experimentation often begin by employing environments that make no attempt to mimic the real world or to control all or even most of the key variables.

This is because of the desire to keep an open mind and rapidly explore large areas of the landscape.

• FIDELITY AND CONTROL IN DISCOVERY EXPERIMENTS

Environments used in discovery experiments are often purposefully artificial in that they provide an over-simplified environment that is designed to focus attention on one or just a few variables. Given that these experiments are focused on what might be possible in the future, it would be very strange indeed if they faithfully reproduced the status quo or reflected the complexities present in the real world. Not having to bear the burden of realism, discovery experiments are relatively inexpensive to conduct. The major costs of these experiments are associated with gaining access to the right set(s) of participants and the instrumentation, observation, and analysis needed to capture what transpired.

Despite the fact that discovery experiments are not overly concerned with realism, their conduct still presents a set of formidable challenges. A careful balance needs to be struck between focusing participants and giving them the freedom to be creative. The lack of control over the values of key variables, which would be a fatal flaw in hypothesis testing experiments, is essential in discovery experiments to foster creativity. Another challenging aspect of discovery experiments is appropriate instrumentation and observation—knowing what to look for and how to capture it. One of the early JFCOM experiments was notable in that the participants expressed a desire to change the way the task at hand was accomplished. Having been given this permission, they reorganized, developed new approaches (doctrine, procedures, and organizational forms), and significantly increased their effectiveness. The experiment was originally designed to get some feel for advanced technology and was well-instrumented for this purpose (quality of information, utilization of system capabilities). Unfortunately, no thought had been given to observing team interactions, the manner in which tasks were allocated or, in fact, the tactics employed. As a consequence, those conducting the experiment could say that the team reorganized, but could not say exactly how (or why). Although the experiment was still enormously useful, its potential value was not realized. Discovery experiments that are designed to observe the characteristics and capabilities of each of the components of a mission capability package are in a much better position to be able to document a new approach and hence make it possible to replicate (or modify, extend) in subsequent experiments.

Good ideas can be generated by participants with appropriate expertise and experience in workshops, seminars, or brainstorming sessions. Other activities can also be good sources of ideas. These include research and/or analysis efforts that draw upon existing research or authoritative historical records, or lessons learned from operations or in other domains. While these less formal approaches to discovery can generate ideas and concepts, they cannot generate evidence that is sufficient to validate definitions of variables, measures and measurement techniques, or patterns of interaction and relationships among the factors of interest. Some systematic experimentation in a more realistic setting is needed to achieve these basic goals.

• FIDELITY AND CONTROL IN MODELING AND SIMULATION

Having identified the portion(s) of the landscape that will be explored during the campaign, the campaign should employ a balanced effort utilizing modeling and simulation as well as hypothesis testing experiments. Using both of these approaches is a good way to collect the empirical data necessary to investigate the selected part of the landscape in greater detail (granularity). Control is an integral part of the use of these tools and techniques. The ability to control the values that key variables take on is needed to be able to focus on particular parts of the landscape (where areas of the landscape are defined by the values of key variables). Control is analogous to the ability to steer a vehicle as it moves over a landscape, while instrumentation is analogous to knowing where you are at any given point in time using, for example, a GPS-enabled map.

Modeling and simulation are often useful to help shape and extend the results of hypothesis testing experiments. Their utility depends upon the validity and granularity of the model or simulation. In other words, experiments are able to represent reality (or future reality) with more fidelity than models, particularly when it involves human behavior, and thus be able to collect meaningful data while models and simulations are able to inexpensively vary the values of variables to represent a wide variety of conditions—something not practical in an experiment.

For physics-dominated problems, there is usually a fairly rich set of data already available. Hence models or simulations can be fruitfully brought to bear when the issues at hand are dominated by physics. However, for issues where human behavior is a major factor, models or simulations built around reasonable sets of alternative behaviors (typically based on prior research or academic literatures) can be helpful in clarifying relationships and eliminating alternative patterns or explanations. These human behavior-centric issues include transformation and coevolution of mission capability packages. Model/simulation results can be used to focus experiments on the areas of greatest potential, uncertainty, or risk.

Increasingly, the biological and social sciences are also generating empirically validated bodies of knowledge that can be used to formulate concepts and hypotheses relevant to individual and organizational behavior. These need to be brought to bear just as the existing physical knowledge is used.

Models and simulations, because they provide a great deal of control over the values of variables, also are easily replicated. But of course, each tool or method has its weakness—for modeling and simulation the comparative weakness is a lack of inherent validity. Their findings are, ultimately, the specific consequences of the set of assumptions built into them that may or may not be reflections of reality. Models will consider only those factors built into them. That is why models and simulations need to work hand-in-hand with empirical experiments or analyses informed by empirical data. Models and simulations are not substitutes for a balanced campaign of experimentation, including human-in-the-loop experiments.

• FIDELITY AND CONTROL IN LABORATORY EXPERIMENTS

As the campaign unfolds, we will normally seek greater and greater degrees of fidelity in the environments being instrumented and observed. Laboratory settings offer an excellent opportunity to observe human behavior at the individual and team levels while simultaneously exercising a substantial degree of control. Thus, laboratory experiments are able to collect meaningful data in the cognitive and social domains.[79]

---

[79] Alberts et al., *Understanding Information Age Warfare*. pp. 57ff.

Laboratory experiments are also cost effective venues for dealing with a number of interacting factors. By introducing these factors in a systematic way into a series of small-scale experiments, their individual and interactive impacts can be assessed before they are introduced into larger exercise contexts where their individual and interactive impacts are very difficult to isolate and assess cost effectively. A series of laboratory experiments is also an excellent way to test the effects of alternative contexts on performance (for example, does a battlespace information system designed for high-end warfare also support counter-insurgency operations?).

• FIDELITY AND CONTROL IN EXERCISES

As we move further up the ladder of fidelity, we begin to sacrifice our ability to control the values of many of the variables of interest. Exercises, for example, having been designed for training purposes, greatly restrict behaviors and alternative approaches, thus preventing the ability of key variables to be controlled (taken on different values in a systematic way). For example, they are not good places to try to introduce future or innovative approaches that have not yet been accepted and incorporated into doctrine. In addition, the results of most exercises are kept closely held to avoid embarrassing individuals who may have made errors as a part of the learning and training process. This restricts the research community's ability to share data and benefit from peer review. The enormous potential value of data from exercises needs to be tapped. To avoid the issues associated with exposing individual performance, exercise data could be sanitized so that while individual identities (or unit identities) are protected, data about the relationships among variables of interest could be used. Moreover, at times some exercise results have been used selectively or out of context to promote or denigrate programs or technologies in inappropriate ways and this has made it difficult to convince those who run exercises to release their data, support experimentation, or to accept an open peer review process to ensure quality control and validation.

However, despite the limited ability from an experimentation control[80] perspective and the other problems associated with exercises, exercises where organizational (and, in some cases where Joint or coalition operations occur, cultural) factors come into play offer environments where ideas can be preliminarily tested in a more realistic setting than is possible in laboratory experiments.

The ultimate in fidelity is to "instrument reality." Instrumenting reality affords an opportunity to systematically record real-world experiences so that data of the requisite complexity and variety can be collected. However, reality provides almost no opportunity for experimental control.[81] Subject selection and experiment design used to control for leadership style, culture (organizational and national), levels of expertise, military schooling, levels of intelligence, and experience in virtual venues is simply not possible in real-world operations, although some statistical controls can be applied if the appropriate data are collected about these factors.

For these reasons, reality is best for confirming something that has been suggested by experiments and as a means of

---

[80] Here, "control" refers to experimental control for the purpose of collecting data under specified conditions and/or to isolate variables and their interactions.

[81] However, if the data collected is rich enough, statistical techniques can be used to create sets of observations that can be used in a similar fashion to those collected in experiments.

serendipitous discovery. Ideas for how one might do something differently or ideas about why something happens are
often extracted from "war stories" (or reports from the field),
after action reports, lessons learned, or analyses of data collected either in the field (often by operations research
analysts deployed with major headquarters or reliability and
maintainability values collected by those responsible for system maintenance) or through surveys of those participating
in specific operations.

The increasing reliance on digital systems in a variety of roles
provides an increasing opportunity for instrumenting reality.
Because data, information, analytic products (intelligence
reports, logistics summaries, plans, etc.), and interactions
among entities are all on the network, the opportunity exists to
capture and archive them for analysis. In at least one case, the
deployment of a "knowledge wall" with a U.S. naval force during Operation Enduring Freedom, a natural experiment
occurred when a new digital capability taken to sea for test and
evaluation was used by the deployed force.[82] With proper
preparation and permission to both analyze the resulting data
and interview personnel, instances like this of instrumenting
reality can yield very rich insights.

Given that military operations, particularly Joint and coalition
operations, are seldom conducted by small groups (there are
some exceptions—some special operations, for example—but
even these efforts are typically linked to larger systems when-
ever network-centric operations or effects-based operations are
contemplated), the network linkages implied are global in
nature. The ability of globally distributed forces and other

---

[82] University of Arizona. Decision Support for U.S. Navy's Combined Task Force
    50 during Enduring Freedom.

Stages of a campaign

individuals and organizations to be on the net is a crucial capability for virtually all future U.S. operations. The rich set of behaviors (information sharing, collaboration, and entity synchronization) required for synergistic efforts means that no force element will operate in isolation. Therefore, it is essential that new ideas, concepts, tools, and ways of conducting operations be assessed in relatively large and diverse settings. Hence, to be relevant, it is very desirable to be able to conduct experiments on the same scale as real-world operations as one approaches the culmination of a campaign. This will mean moving into the very challenging arena of instrumenting reality or re-orienting the nature of large-scale exercises.

PACING

Almost all campaigns of experimentation unfold over a significant amount of time. While it is logically possible to organize, plan, and conduct a campaign of simultaneous or parallel experiments in advance, the results from early experiments often call for significant deviations from plans (not unlike a military campaign). Best practice requires that the conduct of a campaign draw upon the results of other related experiments. The sequence of campaign phases discussed above is designed to facilitate learning and avoid poorly focused, overly ambitious experiments or analyses that have little chance of success. Thus, it must be recognized by those who conduct campaigns that plans for a particular phase cannot be finalized until previous phases are almost complete. Furthermore, one must remain flexible within a phase.

This does not mean that parallel experiments may not be valuable as part of a campaign. There may be situations where experimentation is urgent and/or the opportunity exists to conduct the same or closely related experiments in multiple venues over a short period of time, which effectively precludes these experiments from influencing each other. However, even these opportunities will be richer if the parallel experiments are included in a larger sequential campaign that unfolds over time.

The phasing and timing of specific experiments, analysis, and synthesis activities (incorporating findings into a conceptual model that reflects the state of knowledge) within a campaign is important. Proper timing provides an opportunity for rich analysis of each individual experiment's findings and the opportunity to use these results as inputs to the planning for the other experiments in the campaign or in other campaigns. How much time this requires is a function of (1) the extent to which the experiment design, data collection plan, and data analysis plan are organized to generate clear results promptly and (2) the planning cycle for individual experiments. Efforts in the 1990s (for example within PACOM, the Navy, and the DARPA Command Post of the Future Program) showed that relatively simple, focused experiments could be conducted as frequently as three or four times each year, provided that the processes of data analysis and interpretation were based on detailed collection and assessment plans and received continuous attention. This included focused workshops to review results and design the following experiments. Very simple machine experiments, for example using agent-based models to explore problems by varying assumptions, could be carried out on a more compressed schedule. A series of such experiments would benefit greatly from "peer review workshops" to ensure that their individual results are reviewed thoroughly, alternative interpretations considered, and the results incorporated into the design of the next round of experiments.

Obvious, but worth noting, is the importance of senior peer reviewers (scientific, analytical, and operator) in these efforts.

Increasingly, organizations with long-term responsibilities for experimentation are turning to the idea of "continuous experimentation" in which the same facilities and set of drivers (simulations, scenarios, instrumentation systems, etc.) are employed over and over again to either (1) build indepth knowledge on a single topic or (2) explore a number of topics in the same general arena. While this is an attractive way to reduce the cost of individual experiments, serious effort must be put into ensuring adequate time, effort, and facilities are available to understand and compare the results of individual experiments undertaken using this approach. There has been a tendency to spend the funds and personnel resources for these efforts on hardware and software development and short change research design, analysis, and interpretation. The costs of this misallocation of resources are significant in both wasted effort and loss of knowledge. This is an extension of the argument that "substantial effort is required after the experiment when the results are analyzed, understood, extrapolated, documented, and disseminated."[83]

From the perspective of a campaign of experimentation, this list of activities that deserve adequate funding and attention needs to be expanded to include the design of subsequent experiments in the campaign. The design considerations most likely to be impacted by the results of previous experiments and analyses include the construction of hypotheses (variables and relationships), measurement, controls, scenarios, and subjects. Continuous experimentation can also suffer from a

---

[83] Alberts et al., *Code of Best Practice for Experimentation*. p. 63.

tendency to use the same subjects repeatedly, which can severely limit the generalizability of findings.

Ultimately, the pace of a campaign should be a function of the rate at which key aspects of understanding mature. Maturity of understanding includes not only the ability to predict the behaviors of variables under a full range of conditions, but also includes semantic maturity (clearer and more precise definitions that are understood and accepted throughout the relevant communities), measurement maturity (improvements in measurement concepts as well as measurement techniques and apparatus, reflected in greater validity, reliability, and credibility), more precise formulation of propositions and hypotheses, and broader availability of data and findings for review, critical analysis, and interpretation. The maturity of the conceptual model that underlies a campaign is a good measure of the maturity of the campaign itself. As indicated previously, this can be measured on a scale from (1) vague to precise, (2) isolated elements of knowledge to richly connected explanatory or predictive knowledge, and (3) understandable by specialized researchers to understandable by professional or operational audiences.

EXPERIMENTATION INFRASTRUCTURE

The success of and resources required to conduct campaigns of experimentation depend in large measure on the state of the infrastructure available to support experiments. Therefore, in addition to the best and brightest minds devoted to the design and conduct of experimentation, attention also needs to be focused on creating and/or improving the infrastructure that supports experimentation activities.

This infrastructure consists of (1) people (educated, trained, and experienced researchers and analysts), tools, and venues, all of which are necessary to conduct experiments and analyses, (2) a set of metrics appropriate to the nature of the domain, and (3) mechanisms for disseminating and archiving experimentation and analytical findings and conclusions.

A supportive infrastructure allows experimenters to focus their attention directly on the task at hand rather than on creating the conditions that are needed to accomplish the task. An adequate infrastructure greatly enhances the ability to effectively conduct experiments and analyses, reuse data, leverage lessons learned, and thus increase the yield from experimentation activities and accelerate the rate of progress.

Investments in experimentation infrastructure have been a part of U.S. policy[84] for some time. Government organizations continue to encourage and facilitate scientific advances and associated products by focusing on improving the conditions necessary for success. They do so by funding grants to education, basic research funding, and the creation of dissemination and collaborative mechanisms (e.g., the ARPANET,[85] clearinghouses, and symposia[86]). An assessment of the ability of DoD experimentation infrastructure to support these activities is provided later in the book along with recommendations for necessary improvements.

---

[84] As they are in many countries that promote research and education.

[85] National Museum of American History. "ARPANET."

[86] Significant CCRP efforts are devoted to these kinds of activities, e.g. Command and Control Research and Technology Symposia Series and its Web site.

CHAPTER 6

PLANNING AND EXECUTING CAMPAIGNS OF EXPERIMENTATION

UNDERSTANDING AND ORGANIZING CAMPAIGNS OF EXPERIMENTATION

The *Code of Best Practice for Experimentation* addresses the details of how to conduct individual transformation experiments and introduces the idea of a campaign of experimentation. This chapter carries that discussion further with specific attention on the unique challenges of conducting a campaign. Readers should bear in mind that the key challenges include getting both the theory and its applications right. This involves both building knowledge and developing capability, the twin raisons d'etre for a campaign of experimentation.

Successfully meeting this integration and synthesis challenge makes the difference between merely conducting a series of experiments and actually developing the kind of synergies across individual experimentation activities that are required to contribute to a body of knowledge and ultimately to field innovative capabilities. The myriad challenges associated with conducting successful campaigns of experimentation are underestimated at our peril.

Synergies are needed both to enhance the value of each experimentation activity as well as to develop the broad and deep understanding needed for transformation. Individual experiments and analyses need to be informed by existing knowledge if they are to be successful. These individual activities generate both findings (empirical results and formal analyses of those results) and insights (conjectures about what those findings mean or imply). The analyses that are performed during or immediately following an experiment need to be followed up by other, more detailed analyses or extrapolations involving mathematical models and tools, statistics, and/or simulations.

Within a campaign, of course, individual experiments or analyses are typically linked to other experiments or analyses designed to build on what was learned. Some of the most important learning will be "negative" (learning what is not true and what does not work under the conditions that were represented in the experimental environment). It is sound scientific practice to identify the boundary conditions: the circumstances under which particular findings hold true and the conditions where they break down. Building knowledge requires the integration of all of these pieces of information into a coherent whole. This, as indicated in the *COBP for Experimentation*, requires the use of an overall conceptual model and an understanding of what is not known, as well as the residual uncertainties associated with what is known about the behavior of the variables and the relationships among them.

Communicating among the participants within a campaign, as well as to the larger community, is a key to the success of the campaign at hand as well as to future campaigns. In short, building and applying knowledge is a multi-step, complicated, dynamic, and demanding task. To master it will take commitment and practice. Those undertaking or participating in a campaign would be well-advised to familiarize themselves with best practice and strive to emulate it.

To be successful, campaigns of experimentation need to be not only properly conceived, but also well-executed. While DoD campaigns may differ significantly in their size, scope, and the degree to which they push the frontiers of knowledge, they share to varying degrees a number of attributes that make them very challenging. These include their inherent uncertainty and complexity, the need to create and maintain a culture and environment that encourages innovation, the need to share information across cultural and institutional chasms, the need to balance creativity with pragmatism, the need to push ideas and experimentation activities to their points of failure, the difficulty of measuring and the nonlinearity of progress, and the need for agility as the campaign meets unexpected opportunities, disappointments, and obstacles. This chapter discusses the nature of these formidable challenges and the approaches, methods, and practices that can be employed to overcome them.

While the *COBP for Experimentation* focuses on best practices for the conduct of individual experiments, the focus here is on best practices regarding the design and conduct of a campaign, the relationships among individual experimentation activities that constitute a campaign, and the need for results to be harvested across campaigns. Broadly speaking, the best practices associated with orchestrating individual experimental activities fall into the following broad groups:

- Establishing conditions for success;
- Conducting a sound campaign; and
- Creating a foundation for the future.

ESTABLISHING CONDITIONS FOR SUCCESS

There are two things one can do to create the proper conditions for success: build a strong team and create an explicit conceptual model.

• BUILDING A STRONG TEAM

Although experimentation, even campaigns of experimentation leading to disruptive innovations, has been successfully undertaken in military organizations, individuals with the experience and skills necessary to design and conduct campaigns of experimentation are not currently found in abundance. There is no set of established career paths that is designed to produce the well-balanced teams needed to undertake these campaigns. Thus, the first set of challenges relates to building a team that understands and appreciates the nature of a campaign of experimentation and the requirements associated with individual experimentation activities.

Building a strong team begins with the identification of the experience, skills, and knowledge needed for the conduct of a campaign of experimentation. Having identified the desired attributes of team members, a core team must be established. The core team will need to include experimentation expertise, practical experience, and domain knowledge.

The core team will, in all likelihood, not possess all of the experience, skills, and knowledge needed to conduct a campaign of experimentation. This may be for a number of reasons, for example, a lack of resources or a lack of availability. Thus, the core team will need to be augmented by a number of consultants and/or outsourcing arrangements. An extended discussion of how to build a team is laid out in the *NATO COBP for C2 Assessment*.[87]

The chances of success will be greatly enhanced if team members fully understand the nature of a campaign of experimentation, the specific intent of the campaign at hand, the language and nature of the various kinds of experimentation activities to be employed, the conceptual model and the associated set of metrics that will serve as the avenues of information exchange, as well as the roles and responsibilities of various contributing entities.

• CREATING AN EXPLICIT CONCEPTUAL MODEL

In the previous chapter, we explained both the need for and the nature of a conceptual model. Such a model is the anchor of a campaign and without it the campaign will likely drift. Although the conceptual model will evolve throughout the course of a campaign, it cannot be set to paper and distributed too early because it provides a point of departure for the formulation of the campaign itself.

CONDUCTING A SOUND CAMPAIGN

Conducting a sound campaign requires quality control in the form of peer reviews, sound documentation practices, keeping

---

[87] *NATO Code of Best Practice for C2 Assessment.* pp. 31-36.

the conceptual model up-to-date, and conducting analyses beyond individual experiments.

• PEER REVIEWS

While it is a given that each individual experiment should reflect high quality analyses and peer reviews to ensure that they are placed in the context of existing knowledge, this attention to individual experiments is not a substitute for proper attention at the campaign level of management. Those conducting campaigns of experimentation have an obligation to maintain a peer review system that cuts across individual experiments and analyses to ensure that findings and the interpretation of those findings are continually brought together, along with other new research and literature, to ensure knowledge development.

Peer review is the most powerful tool those conducting a transformational campaign has in his or her arsenal. While the team that performs these campaign-level reviews should include individuals who also act as peer reviewers for individual experiments, it is important that this campaign-level team includes some senior reviewers who are not directly associated with individual experiments or analyses. The need for outside expertise is based on the need for review at the knowledge base level, which requires peer reviewers with broad and deep expertise that extends across the whole range of relevant substantive areas, across all levels of experimentation setting fidelity, and also extends from discovery through hypothesis testing to demonstration experiments.

Transformation presents a particularly challenging problem for the peer review process. The basic idea, of course, is to get knowledgeable and experienced individuals who can provide objective insights and counsel. Given that transformation is about disruptive ideas, turning to the same people who are comfortable with and experienced with the current ways of doing things does not make a lot of sense. Given that DoD transformation is about the future, a future that is arguably very different from the present, there is arguably no one with the requisite experience. So who should be the "peers"? Under these circumstances, the answer lies in assembling a diverse team. This means including individuals from other domains who may, for example, be experienced with new networked organizational forms or individuals from organizations that need to work with the military in civil-military operations. Many of the right individuals may not be well-versed in military operations. It is the team *mix* that counts. Having the right mix of perspectives is what is important, not trying to find it all in one or two people and hence sacrificing important perspectives to established military credentials.

• DOCUMENTATION AND KEEPING THE CONCEPTUAL MODEL UP-TO-DATE

Each of the individual experiments or analyses will generate a variety of raw materials and products (e.g., data, planning documents, archives of the experimentation artifacts, databases, reports, and briefings). The quality of these products is crucial to building knowledge. Thus, it is important that they have been peer reviewed systematically from the earliest plans to the final reports.

For purposes of building knowledge, the most important elements are (1) consistent language (clear and operational definitions and measures), (2) explicit use of metatags (metadata) on data, and (3) clear and complete descriptions of assumptions. These are part and parcel of an explicit conceptual model.

All three are essential to ensure that data and knowledge can be reused by others both within the campaign and in other experiments and campaigns. Consistent language makes it possible for others to measure the same thing in the same way. The metatags provide additional context and, together with consistent language, are the crucial elements needed to link the empirical results of the different experiments. They are also the insurance system that prevents unlike things, such as apples and oranges, from being inappropriately mixed. Both language and metatags need to exist prior to the design of a campaign, or the individual experiments, or analyses that comprise the campaign. If they do not, an initial version will have to be developed in the formulation phase of the campaign and fleshed out and refined in experiments and analyses over the course of the campaign. The former is definitely preferred to the latter, but a complete set of definitions and metatags is seldom possible prior to the conduct of the first experiments and analyses. In any event, these definitions, associated metrics, and metatags will need to be updated as the campaign unfolds. Finally, it is important to document the assumptions because without explicit treatment, the "all things being equal" condition will be met only by absence, introducing noise into the experimentation process.

A database is required to organize and make raw data and other empirical results from experiments and analyses accessible so that the state of knowledge as it relates to a conceptual model can be understood. In an experimentation campaign, having a database-driven conceptual model keeps the conceptual framework up-to-date and provides the basis for the design of individual experiments and analyses by providing an up-to-date view of what is known and not known.

To accomplish this, beginning at the launch of the effort, someone needs to be responsible for updating and maintaining the conceptual model and its underlying database with operational definitions and metatags. This role will be necessary throughout the campaign, both as a cross-check on the work of the experimenters and as an integrating function as results become available. This is not a passive role. The processes of experiment design and execution will, almost undoubtedly, reveal new factors that must be included and new categories of metadata that must be tracked. Hence, maintenance of both the conceptual model and the structure and substance of the database will be a continuing challenge throughout the campaign. Data, while providing the basis for knowledge, are not in and of themselves knowledge. Thus, the objective of a campaign of experimentation is more than the collection and categorization of data. The objective is an integrated knowledge base—a rich depiction of the existing knowledge, gaps in knowledge, uncertainties, cause-and-effect relationships, temporal dynamics, and underlying structures.

This task is both a science and an art, and should be the responsibility of the senior technical person managing the campaign. An instantiated conceptual model includes an integrated database. To help individuals visualize the overall structure of the model and the relationships in detail, both graphical and textual materials are needed.

The conceptual model and various views or perspectives of it have a diverse set of audiences. Different views will be appropriate for each of these audiences. The underlying collection of data will obviously need to be updated regularly as results from individual experiments become available and are integrated into the larger picture. Keeping it current will also provide those managing the campaign with the capacity to identify (1) new opportunities, (2) ambiguous findings that will require more work, and (3) important findings requiring fundamental adjustments to the existing theories and knowledge.

In DoD, the vital activities associated with developing and maintaining the conceptual model have often been starved for resources and management attention. This appears to be a reflection of the mistaken assumption that individual experiments are the determinants of success. As we have previously stated, while individual experiments must be well-designed and executed, they do not effectively stand on their own. Rather, they are inputs to the larger process of building a body of knowledge and experience.

Hence, the integrating functions of a conceptual model must be seen as vitally important, not "extra duties as assigned" that will be easily handled by the leadership of the campaign. As such, they need to be appropriately resourced.

• ANALYSES BEYOND INDIVIDUAL EXPERIMENTS

A campaign of transformational experimentation must be able to make direct comparisons between and among findings and insights from the individual experiments in the context of the existing body of knowledge. This kind of analysis capability must be built into the team and appropriate analytic activities need to be planned into the campaign as it is being designed. Simple pair-wise comparisons will be essential, but the real gains in knowledge will be made by discovering patterns that emerge by comparing findings from several experiments and by analysis of an updated conceptual model.

Knowledge is "lumpy" just like reality and seldom takes the form of simple linear relationships or unconditioned conclusions. This is almost always the case when serious questions are being addressed. Hence, extending the focus of analyses beyond the individual experiments is a necessary part of knowledge building. In many cases, this will require creating simple simulations or models that focus on filling in the gaps. Resources need to be allocated to this type of work. Experimentation teams working on individual experiments need to participate in this effort because they are familiar with key data and the conditions under which it was collected, and are supposed to be prepared to conduct such analyses after each experiment.[88] However, the campaign manager will also want to have additional resources available for cross-cutting analyses in order to be able to update the conceptual model and modify campaign plans and experiments. These extended cross-cutting analyses (note that they are not new experiments unless the campaign is using modeling or simulation experiments as an important tool) should be ongoing and be reassessed as soon as possible after the completion of each individual experiment. This places a premium on the plans for individual experiments being realistic enough so that the availability of findings and insights is predictable. In some cases, these analyses can be conducted in parallel with peer reviews in order to save time; however, the richer set of findings and insights emerging from these extended analyses also require peer review at the level of the campaign rather than at the level of the individual experiment.

---

[88] Alberts et al., *Code of Best Practice for Experimentation.* p. 317ff.

A major purpose of these extended analyses is to enable campaign managers to continually assess (and when necessary, update) the campaign plan. In many cases the ability to take advantage of knowledge breakthroughs or to revisit issues where the experiment results are either ambiguous or are inconsistent with those expected or hypothesized will determine the long-term importance or success of a campaign. The pace of the campaign will in large part be a reflection of these integrative analyses. Rushing ahead with additional experiments without taking the time to perform these extended analyses will often prove to be a major error and result in experiments that are far less valuable then they could have been.

In some cases where real urgency has been perceived, thoughtful campaign managers have scheduled the extended analyses in parallel with the core experiments and aimed the analyses at downstream experiments rather than slowing the overall process.

CREATING A FOUNDATION FOR THE FUTURE

As with any scientific or knowledge building endeavor, the richness of a campaign of experimentation (and its impact) depends on the extent to which results are disseminated. The forums of the products and information to be disseminated should include peer reviews, presentations and models, workshops, and professional journals.

The peer review process is a potentially powerful contributor to effective dissemination to the extent that reviewers are drawn from a variety of domains and the extent to which reviewers are looked to as thought leaders. Peer reviews will sow seeds that will eventually mature across a wide landscape. Senior reviewers with rich professional networks provide multiple channels for broad and deep dissemination of products and results across multiple domains. Peer reviewers should be encouraged to pass results along to those they respect. Similarly, the research teams themselves should be encouraged to disseminate their products. This best practice stands in stark contrast to misguided efforts to control the distribution of products in an effort to ensure that they will not or cannot be misunderstood. The notion of information ownership, whether the information source is a sensor or an experiment, is antithetical to the tenets of NCW and contrary to current DoD information policies that demand interoperability and widespread sharing of information.

If experimentation in DoD is to contribute to our efforts to transform, then experimental results must be clearly articulated and findings reported completely and honestly. The products of an experimentation campaign must carefully distinguish between findings (what the data and the patterns in the data are) and insights (what the researchers believe they mean or imply). Confident, quality research teams develop products they are proud to distribute and welcome honest scholarly feedback.

In addition to reports documenting research findings, campaigns of transformational experimentation will also produce presentations (briefings, videos, constructive models, etc.) that are interesting and important. These materials should also receive widespread dissemination across the relevant research and policy communities. Outreach is crucial if experiments are to have broad impact. Widespread dissemination will also assist communities of interest in building knowledge and reducing the amount of redundant or less than well-focused research being conducted. Indeed, broad dissemination will often influence other research agendas and lead to related research that builds on what has been done. The likelihood of this cross-fertilization will be much greater if the conceptual model with its supporting integrated database and knowledge base are well-documented and are also made available. This allows other research teams both to understand what has been done and also to build directly on it. Obviously, an outreach effort to disseminate knowledge will include materials tailored to different audiences including policy makers, warfighters, acquisition officials, and researchers. The knowledge base should have been constructed and maintained with these different types of audiences in mind—this should not represent a major challenge.

Campaigns of experimentation should also include among their activities the conduct of meaningful professional workshops designed to ensure cross-fertilization across the individual experiments and among campaigns and other related research and development efforts. Some campaigns will find it useful to design and conduct their own workshops. Some may want to have their own conferences or symposia. All of them can and should be looking for professional meetings where their efforts can be discussed and understood. The Command and Control Research and Technology Symposia (CCRTS) series provides an internationally recognized venue for papers on most transformational subjects and posts these papers on its Web site[89] where a larger community can access and download them. Professional societies such as the Military Operations Research Society (MORS), the Armed Forces

---

[89] www.dodccrp.org. The CCRP Web site receives over 5 million hits every year.

Communications and Electronics Association (AFCEA), the National Defense Industrial Association (NDIA), and the Institute of Electrical and Electronics Engineers (IEEE) all routinely hold major meetings where appropriate results can be presented.

Finally, publications in professional journals are also appropriate mechanisms for communicating with the broader research communities in both academia and industry. While slow because of their editorial processes, many of these refereed journals also carry both prestige (because of the peer review processes involved) and broad impact. These journals are also increasingly available on the Web, which improves their reach.

TYPES OF TRANSFORMATIONAL CAMPAIGNS

While there is quite an extensive body of literature that discusses experimentation in both the physical sciences and in the social sciences,[90] there has been very little written on the planning and execution of campaigns of experimentation. One of the most widely cited texts on experimentation[91] devotes a total of three pages to what they term "Multi-study Programs of Research," the term they use for what we refer to as *campaigns of experimentation*. While brief, their discussion is useful. It notes that,

> Multi-study programs of research offer great control over exactly which aspects of generalized causal inference are examined from study to study, so that the researcher can pursue precisely the questions that

---

[90] Wheeler and Ganji, *Introduction to Engineering Experimentation*. 2003.
   Campbell and Russo, *Social Experimentation*. 1998.
[91] Shadish et al., *Experimental and Quasi-Experimental Designs*. 2002.

most need answering at a given moment in time. However, these programs also require much more time and funding than any other method we cover.

The other methods they refer to include individual experiments and quasi-experimental designs.

They distinguish between two types of multi-study programs: (1) Directed Programs of Experiments and (2) Phased Models of Increasingly Generalizable Studies.

- *Directed Programs of Experiments* systematically integrate over many experiments the explanatory variables that may account for an effect. These can be thought of as a structured effort to extend the number of trials for a single experiment in which alternative causes, ways of measuring the effects, and sets of intervening variables are introduced. This effort to integrate over a number of individual experiments is not unlike the "distillations" that Project Albert and other agent-based modeling efforts use to improve their understanding of observed phenomena. As such, these efforts are relatively simple to plan and execute because they can be thought of as one very large experiment that continues over time or is carried out by different teams of researchers at the same time.

- *Phased Models of Increasingly Generalizable Studies*, on the other hand, proceed from basic research through organized phases until the research focuses on applications. Shadish et al. cite practices in the National Institutes of Health (NIH) and the Food and Drug Administration (FDA) in which the first effort is in basic research to identify promising approaches; the second focuses on identifying testable hypotheses, including developing valid and reliable ways to measure the phenomenon of interest; a third takes the form of intervention trials to test the alternative solutions; and the fourth takes the form of effectiveness studies to determine the impact of the intervention on real-world populations.

While DoD may employ both of these types of multi-study programs in its campaigns of experimentation aimed at transformation, this book does not discuss "directed programs of experimentation" because their planning and execution is not radically different from what is necessary to plan and execute a single large experiment. This subject is adequately covered in the *COBP for Experimentation.*

Moreover, the most important transformational experimentation within DoD, and the most difficult to carry out successfully, are concept-based campaigns that proceed in phases, if not to a generalizable solution, then to one that can be implemented. As is discussed in some depth below, these phased campaigns of experimentation are more difficult because the findings from the early phases have a profound— and potentially unpredictable—impact on the planning and execution of the later phases. Hence, the teams involved and the approaches they take must be far more agile than those supporting directed programs of experimentation. Moreover, the transitions between phases are often difficult, in no small part because they typically involve different groups of people with different charters and standards for success.

Both JFCOM and the Services tend to follow a phased approach along the lines of the process that is described in *Understanding Information Age Warfare*[92] and *Information Age Trans-*

---

[92] Alberts et al., *Understanding Information Age Warfare.* pp. 50-51.

formation[93] to coevolve mission capability packages. For example, JFCOM uses Project Alpha, a small team of selected professionals (often supported by universities and other research organizations) to explore promising ideas and technologies. This work often takes the form of discovery experiments. Their results, if promising, are turned over to a concept development team, which undertakes both discovery and some limited hypothesis testing (primarily hypothesis refinement) experiments. The ideas that mature into promising concepts are then turned over to another team for prototype development, which for the most part involves hypothesis testing and demonstration experiments. The best results from these efforts are integrated into large Joint exercises and experiments, and then become the ideas and technologies that the Joint Staff and the Combatant Commands introduce into the field. This theoretical progression is clearly a phased model. It is a theoretical progression because concepts and ideas do "jump the tracks" at times—very promising new ideas may be pulled forward into rapid prototypes if they appear to deal with an urgent need, some ideas are found wanting and move backward in the process so they can be improved, and of course, some ideas "die" or are eliminated in the process.

MCP MATURITY

Much of DoD experimentation is focused on delivering specific mission-related capabilities. Only a small fraction of DoD experimentation is conducted to build the body of knowledge that is needed to underpin efforts at transformation. Getting the theory right cannot be overstressed. Without a basis in theory, efforts at coevolution may proceed on random paths.

---

[93] Alberts, *Information Age Transformation.* pp. 75-77.

Campaigns all proceed from a set of immature ideas (in the form of either a concept for a MCP or a theory) and as they move from one phase to the next, develop increasingly mature mission capability packages that coevolve all of the elements needed for a fielded innovation—organization, command arrangements, training, weapons systems, logistics, leadership, doctrine, tactics, techniques, procedures, C4ISR systems, and so forth and/or a better understanding of the theory that underpins a transformational concept or capability.[94] As shown in Figure 4, the maturity of a MCP-focused campaign of experimentation can be seen on at least three levels:

- As the MCP (or knowledge) matures, the purpose of the relevant experiments in the campaign shifts from discovery to hypothesis testing to demonstration.

- The process is one that builds on existing knowledge (research reports, books, lessons learned, journal articles, etc.) and the available relevant expertise. Over time, a successful campaign of experimentation both influences existing knowledge and available expertise and also comes to enhance and extend the knowledge on the particular topic of interest.

- The MCP (or our effort to better understand something) begins as a set of ideas (variables and concepts) and matures to include explicit information about relationships between those ideas, the performance of the systems they involve, and their effectiveness (impact on the operating environment), and finally to the limiting conditions (the conditions necessary for success or the conditions under which the relationships do not obtain); ultimately resulting in an implementable MCP (a fielded

---

[94] Alberts, *Information Age Transformation*. pp. 75-77.

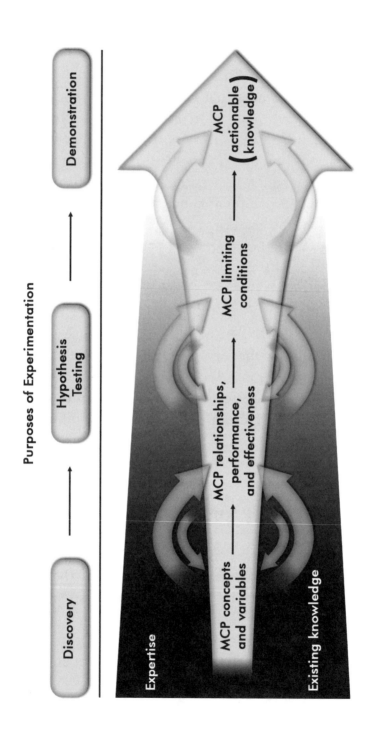

Figure 4. Maturity of MCP/Theory

MCP maturity

capability) and/or actionable knowledge that contributes to our understanding of the theory.

These perspectives each tell us something about the progress that is being made.

PHASES OF A CAMPAIGN

Clearly, successful campaigns of experimentation will go through a number of distinct phases on the road to creating mission capability packages, producing actionable knowledge, or contributing to theory.[95] We think it useful to think about a campaign as having four phases: formulation, concept definition, refinement, and demonstration.

FORMULATION

Campaigns all begin with an idea. Sometimes the idea is ill-formed and sometimes it is precisely expressed. The idea may come from anywhere: experience in new types of battlespaces or operations;[96] new theories of how to operate;[97] literatures borrowing from new areas of inquiry;[98] formal requirements such as the imperative to digitize U.S. Army operations;[99]

---

[95] In the discussion that follows we focus on campaigns aimed at developing MCPs. These phases also apply to campaigns that develop theory.

[96] Alberts and Hayes, *Command Arrangements*. 1995.
Wentz, *Lessons from Bosnia: The IFOR Experience*. 1999.
Wentz, *Lessons from Kosovo: The KFOR Experience*. 2002.

[97] Macgregor, *Breaking the Phalanx*. 1997.
McNaugher et al., *Agility by a Different Measure*. 2000.

[98] Moffat, *Complexity Theory*. p. 156.
Office of Force Transformation. "Operational Sense." 2004. p. 7, 10, 14.

[99] Alberts et al., *Understanding Information Age Warfare*. pp. 254-261, 265-269.
"U.S. Army Training and Doctrine Command." Army Transformation Wargame 2001.

technological opportunities;[100] or ideas derived from conjunctions of all of these types of initiatives.[101]

The formulation phase of a campaign of experimentation serves to establish what the campaign is all about, its specific objectives, participants, and the resources that will, at least initially, be required.

CONCEPT DEFINITION

Regardless of their origin, campaigns begin as ideas that are expressed in general terms during the formulation phase. However, ideas cannot be productively pursued until they are expressed in a coherent and explicit manner. Over time, scientists and analysts have found that developing an explicit conceptual model is an essential first step to create a common understanding of the point of departure and a focus for ongoing efforts. Conceptual models consist of relatively simple elements (ideally primitives) that are linked together to form relationships. The original NCW tenets constitute such a conceptual model—a model that is very simple in its expression, but rich in implications.

Very often, an initial version of a conceptual model is not expressed in the form of testable hypotheses, either because the language involved is not precise enough or because some of the concepts are abstractions that cannot be directly observed and measured. Moreover, the assumptions underly-

---

[100] RAND. NCO Conceptual Framework Case Study Template Illustrative Example: Adapted from RAND Air-to-Air Case Study.

[101] Alberts et al., *Network Centric Warfare*. 1999.
Network Centric Warfare Department of Defense Report to Congress. 2001.
Alberts and Hayes. *Power to the Edge*. 2003.

ing the conceptual model and the limiting conditions (circumstances under which the posited relationships will not occur) are seldom fully understood or expressed in the initial conceptual model. During this phase, a conceptual model must move beyond such an initial formulation and develop the precise operational definitions necessary to test the relationships that constitute the model.

The NCW tenets, for example, include a number of terms that have taken some time to define precisely and measure accurately. The concept of *shared awareness* is an excellent example of an important variable that has required considerable research both to define clearly and to measure adequately. Because awareness resides in the cognitive domain (within people's heads), it is difficult to measure. Shared awareness is a construct derived from the characteristics of the state of awareness of a set of individuals. To measure or calculate shared awareness requires a mechanism that can compare and contrast the cognition of more than one individual. Needless to say, while efforts have been made to measure both awareness and the degree to which it is shared among a group of individuals/organizations, much more work is needed to fully explore this concept.[102]

The concept of "self-synchronization" is another excellent example of a concept that is difficult to operationalize. This is an idea for which the identification and specification of the underlying assumptions (competence of the force, trust in one another, etc.) and limiting conditions have taken some time to develop and understand well enough to measure.

---

[102] See: Kirzl et al., "Command Performance Assessment System." 2003.
Alberts et al., *Understanding Information Age Warfare*. pp. 239-282.

Fleshing out the conceptual model is a necessary prerequisite for moving on to the next phase. For campaigns of experimentation that are focused on the development of MCPs, this is where the elements of a MCP take shape. This is where the nature of the organizational, doctrinal, materiel, leadership, training, and other elements of the MCP concept take form and provide the initial point of departure for the process of coevolution. This phase (concept development) consists of experiments and analyses aimed at discovery, during which (1) the initial concepts are enriched, (2) the concepts are turned into falsifiable hypotheses, and (3) measurement tools and apparatuses are developed to allow systematic examination of them.

REFINEMENT

Once a concept for a MCP is expressed in suitable detail, it is ready to undergo rigorous testing (and be refined accordingly). At this point, the campaign can move ahead with a series of hypothesis testing experiments. These experiments need to alternately test both the concept itself and how it is being applied. This may seem obvious; however, some in DoD have confused the two, mistaking the testing of an application for the assessment of the concept behind it. The refinement phase is the heart of a campaign of experimentation and all too often is not as rich or as rigorous as it should be.

Because of the rich problem space in which DoD missions now occur, including warfighting, surveillance missions, peace operations, missions as security forces, support for police functions, and support to civilian authorities in humanitarian relief,[103]

---

[103] Alberts et al., *Code of Best Practice for Experimentation.* p. 202.

this phase of a transformational campaign of experimentation needs to be broad, incorporating a variety of settings as well as transitions between and among different parts of the mission space. Adversaries may range from nation states to transnational organizations (e.g., drug cartels, businesses engaged in illegal arms transfers, or organizations clandestinely trading in the materials needed to develop weapons of mass destruction), sub-national organizations, terrorist networks, computer hackers, or even disturbed individuals capable of harming large numbers of people.

If allowed to concentrate on missions of a particular type, the results of the experimentation will tend to optimize on the performance of that specific mission. As a result, we may lack the agility needed in twenty-first century U.S. and coalition forces. Some systematic sampling of the relevant mission space[104] will be needed to keep this phase manageable while also ensuring that enough richness is present to ensure that agility is considered.

In addition to a need to address the full range of potential missions and circumstances that comprise the mission space necessary to ensure force agility, the conduct of this phase needs to be rigorous. Rigor has not been adequately stressed to date in the Department of Defense. One possible explanation that has emerged from discussions with those engaged in DoD experimentation is that there is a reluctance to conduct rigorous experiments because of the imperative to be "successful." DoD organizations have a tradition of defining success as the success of the concept, not the success of the experiment. This is simply counter-productive. However, DoD reward systems encourage this dysfunctional behavior. There is a tradition of

---

[104] *NATO Code of Best Practice for C2 Assessment.* pp. 163-186.

judging DoD Program Managers by the "success" of their programs (adoption and institutionalization). This makes them
advocates for ideas and systems rather than guardians of the
experimentation process. Thus, "bad" ideas tend to be protected rather than discarded and the process of MCP
development suffers. What needs to be understood is that *success* here is all about learning and adding to the body of
knowledge in useful ways. Learning what does not work or the
conditions under which something breaks down is very valuable and is a *success* despite the fact that the specific treatment
in the experiment was not. This will ultimately result in a
greater rate of progress in knowledge acquisition and more
capable MCPs. Business as usual has and will continue to
result in less capability and a greater risk of failure.

This is also a challenge when the entire incentive system
(from Congress to DoD executives) is focused on short-term
returns on investments. The history of disruptive innovation
points to the fact that in the early stages of the development of
a disruptive idea or capability, the concept itself may be
strong, but its implementation not sufficiently matured to be
competitive.[105] For a novel idea, some portions of the early or
initial versions of the conceptual model are very likely to be
wrong. Hence experiments can be expected to identify new
variables and new relationships that must be accounted for in
the theory and will result in modifications to the conceptual
model. This is natural. Discoveries of this sort are not failures;
they are successes.

---

[105] See: Keegan, *The Price of Admiralty.* pp. 183-190;
    Corum, *The Roots of Blitzkrieg.* 1994.
    Christensen, *The Innovator's Dilemma.* 1997.

Deciding when to abandon a concept for a MCP or an implementation of one, when to loop back and reformulate, when to remove elements, when to make minor changes, when to merge elements of one MCP with others to gain synergy between them, and when to peel off some valuable and promising elements and accelerate their development are among the most difficult choices to be made in the course of experimentation and the conduct of research and development. Stubbornly moving straight ahead is seldom the right approach, but the alternatives require both careful thought and also a rich understanding of what is happening in related fields of inquiry and innovation.

As both the concept for the MCP and its instantiation mature, hypothesis testing experiments move into arenas where the concept of effectiveness[106] becomes key, the applications arenas become narrower, and the limiting conditions associated with a MCP will emerge. Success here is finding the limits. Therefore, some experiments must result in the failure of an application under a certain set of conditions. Those involved in DoD experimentation need to fully understand this or MCPs and concepts for MCPs will never be adequately tested.

Here again, it is important to explore a range of missions and the transitions between missions. The users, those who will live with the consequences of adopting the innovation or MCP, must now play a more important role, including their involvement in the development of the appropriate procedures (e.g., doctrine or TTPs), organizational forms, and training associated with success.

---

[106] Desirable impact on the operating environment or mission accomplishment in terms of effects-based operations—i.e., the intended effects with a minimum of negative unintended consequences.

DEMONSTRATION

Once these organizational concepts and processes are established, demonstration experiments (where a value of a known capability is demonstrated in a realistic context) take center stage. Success at this level is a close precursor to broad adoption of the innovation and should enable a relatively rich and easy transition into the force. DoD has become accustomed to demonstrations as a way of showcasing mature technologies and bringing them to the attention of the user community. As a result, there is very real danger of moving to this phase too early—before the necessary rigorous testing has been done and the MCP is mature enough.

GUIDING PRINCIPLES FOR CAMPAIGNS OF EXPERIMENTATION

The principles that guide campaigns of experimentation are:

- Recognizing the importance of phase transitions. Continuity in personnel is important so that the ideas do not become distorted during transitions while an idea moves from a conceptual model to testable hypotheses, to the exploration of limiting conditions, to effectiveness testing and implementation as a MCP.

- Recognizing that the process as well as the progress made are nonlinear. A host of decisions will need to be made about what to retain, what to change, and when to make phase transitions. Learning throughout the experimentation process may cause abrupt changes (either a need to revisit or a breakthrough in understanding) in the pace of maturation.

- Making resources (key personnel, funds, access to facilities, etc.) available on a continuous as well as flexible basis. Brittle planning will cause unnecessary disruptions and delays. Artificial barriers will break momentum.

- Recognizing the importance of senior steering groups with the authority to make decisions about the direction and pace of the campaign, the allocation of resources, as well as assignment of key personnel.

- Recognizing that a campaign of experimentation needs to look at a wide range of application situations: warfighting as well as support to civilian leadership, and coalition as well as Joint.

CHAPTER 7

EXPERIMENTATION CAMPAIGNS: STATE OF THE PRACTICE

INTRODUCTION

This code of best practice is unlike traditional codes of best practice in that it addresses an emerging competency rather than an area with a well-developed theoretical foundation and a mature practice. Thus, this COBP will conclude with a chapter that lays out "a way ahead," identifying the actions needed to improve the state of the practice. To set the stage for this discussion, an assessment of current, planned, and past attempts to conduct campaigns of experimentation is presented here.

When this volume was conceived, the authors had hoped to identify several successful campaigns of experimentation to use as illustrative examples. However, a review of DoD campaigns undertaken to date found that (1) most of them were far from exemplary and (2) these efforts encountered a variety of systemic obstacles related to the practices[107] of the Department of Defense.

This is not to say that there have not been good, even excellent, experimentation activities conducted by DoD. There most certainly have been high quality research efforts, but these have been what Shadish et al. would term "directed programs of experiments." Furthermore, these efforts have been almost always too narrowly focused and undertaken too early in the development cycle to support the coevolution of coherent mission capability packages. While they have often improved the state of knowledge on a particular topic, that gain in knowledge has seldom been successfully exploited and carried into realistic enough contexts that would allow one to validate the results, and determine their mission potential or impact across the mission spectrum.

A simple story illustrates the problems with transitioning the knowledge and experience gained in these experimentation activities into real capabilities. A number of years ago, DARPA (Defense Advanced Research Projects Agency) was completing its work on one of its most profound and important technical innovations—packet switching. The technology had been developed and demonstrated successfully and was very much the "talk of the town" in the research community. The next logical step in the process (of transition/MCP development) was to organize and conduct experiments to assess the potential impact of this new technology on military operations. This effort would include the development of a basic research design, including key metrics, and the conduct of a number of experiments. Unfortunately, there was a corporate culture in

---

[107] These include, for example, how we carry out analysis, R&D, doctrine development, training, and test and evaluation.

DARPA at the time that believed DARPA's role was to develop new technologies, not to assess their impact or take the responsibility for transitioning them into the force. No one in DoD stood up to accept this as their responsibility. This resulted in a "transition strategy" that, as expressed by some long-time DARPA hands, was to "develop something new and potentially important, make it known to the defense community, and wait for it to show up in proposals in a couple of years."

More than a decade later, DARPA was developing a prototype war room intended to examine how some of its products could be integrated into something resembling a "system of systems" and transitioned into DoD and the force. This "wait and see" approach to technology insertion has been replaced in recent years by the creation of a DARPA/DISA "Joint Program Office" intended to improve transitions and efforts to work more closely with the Services. However, efforts such as this remain the exception, not the rule.

While these examples are drawn from DARPA, the experience of that organization is not atypical of the organized research community within DoD, and the existence of the transition problem is still widely acknowledged. Transitioning to new technologies and capabilities has been a challenge for many years. When the change is "disruptive," transition is a more formidable challenge that should be addressed through campaigns of experimentation. This is because the classic system of writing requirements, staffing them against current priorities and needs, having committees of various sorts vet them, and building to the resulting requirements seldom results in disruptive innovation. JFCOM has been given some relevant charter and is working to organize their efforts to move from good ideas to mature concepts, to useful prototypes, and then into the force, but their early efforts have also met with mixed results.

This chapter looks at the current practices by which DoD carries out campaigns of experimentation, then reports on some "misadventures in campaigns" that have occurred during the last decade, and concludes with a discussion of some of the self-created barriers that make if difficult for DoD to organize successful experimentation campaigns that promote disruption innovation.

TYPICAL CAMPAIGN PRACTICES

There are five types of organizations within DoD where campaigns of transformational experimentation are likely to occur:

- Basic research organizations;
- DARPA;
- The Services;
- Combatant and Functional Commands; and
- Joint Forces Command.

The basic research organizations are chartered to develop new knowledge in areas where the Department of Defense will benefit. Their work tends to focus on issues where universities, very small companies, and in-house researchers dominate, in no small part because the resources available are small and novelty is valued. The Office of Naval Research (ONR), the Army's Research Laboratory and Research Institute (ARL and ARI), and the Air Force Human Resource Laboratory (AFHRL) are good examples of these types of organizations. They are often successful in building some new knowledge by focusing resources on a narrow topic and pursuing a multi-year program. A very successful effort, such as ONR's TADMUS (Tactical Decision Making Under Stress) Program of the 1990s, produces academic papers and doctoral dissertations, journal articles, conference and symposium papers, and a cadre of experts who can carry the new knowledge into the operational community.[108] These organizations do work that is clearly more similar to directed programs of experiments than to transformational experimentation. However, their efforts are often valuable to later efforts to build mission capability packages.

DARPA is chartered with undertaking high-risk, high-payoff research. Its efforts frequently extend over 5 years, providing an excellent venue for developing campaigns of experimentation. DARPA is dominated by people with backgrounds in the "hard" sciences and engineering. It long ago stopped working on social and behavioral science issues and only in the past few years has focused serious efforts on systems development. DARPA campaigns are seldom driven by experimentation. They tend to focus, instead, on developing and demonstrating technologies. The focus of the effort is increasingly to identify one or more transition targets: a Service, JFCOM, a Combatant Command, or a Functional Command. This allows DARPA to focus its resources on research and development while planning for a hand-off to the user community. As a consequence, a DARPA program does not generate a mission capability package in itself, but is more likely to provide the

---

[108] Various TADMUS documents can be found at the program's Web site: http://www-tadmus.nosc.mil/viewdocs.html. (Aug 2004).
These documents include papers presented at the Command and Control Research and Technology Symposia (1996-1999), the IRIS National Symposium on Sensor and Data Fusion (1997), and the 40th Human Factors and Ergonomics Society Annual Meeting (1996).

materiel part of a DOTMLPF package that is a possible input to a campaign of experimentation that is conducted by the transition target. As this process matures, however, those transition partners are becoming more involved in the early stages of DARPA programs, encouraging richer coevolution.

The Services, with the largest set of resources, have an opportunity to carry out a variety of experimentation campaigns. They include the basic research organizations discussed earlier as well as a variety of organizations that are capable of applied research and development. For example, the Army includes a number of battle laboratories, Communications-Electronics Command (CECOM), Concepts Analysis Agency (CAA), and TRADOC Analysis Command (TRAC), all with very real research capabilities, and can call on the RAND Corporation for FFRDC support. The Navy has Space and Naval Warfare (SPAWAR), the Naval Postgraduate School (whose faculty is actively involved in a variety of research), the Marine Corps Combat Development Center (MCCDC) and the Center for Naval Analyses (CNA), as well as a variety of organizations that do R&D for particular types of naval platforms. The Air Force has its Electronic Systems Division (ESD), Rome Air Development Center support from the Mitre Corporation, and other specialty organizations internally, but relies heavily on major contractors for R&D on new systems and platforms.

Efforts at Army digitization have led to the identification of units where experimentation was a specific responsibility. However, the pressure arising from operational tempo has made this impractical in recent years, so the development of new types of force capabilities, such as Stryker Brigade, and data collected from applications in the field (for example, the Command Post of the Future DARPA technology in Iraq) have become key drivers for innovation and change. Hence, the Services have a plethora of organizations and institutions that can carry out campaigns of experimentation. Some excellent work gets done within them. For example, MCCDC routinely focuses its efforts on specific experiments considered important by the Marine Corps. Some of the Army's Battle Laboratories have done focused work on difficult issues like urban combat and training for it. However, the vast majority of the R&D work in the Services employs a model that focuses on technology first, then later looks at building mission capability packages. Almost all of this work goes on in a single Service context, which minimizes opportunities for Joint and inter-agency experimentation. A positive development has been the willingness of Services to experiment in cooperation with close allies (notably Canada, the U.K., and Australia), who often seek well-crafted campaign designs to ensure their limited funds are well-spent.

Special Operations Command (SOCOM), which has legal authority comparable to that of the Services, has also become a hub for innovation and some experimentation. It may benefit from its small size, heavy operational responsibilities, and recruiting selected personnel, however, its activities are largely shielded from public view.

Combatant and Functional Commands are dominated by relatively short-term issues. However, they often recognize that they face challenges that require innovation, which makes them willing to engage in campaigns of experimentation. For example, when Admiral Blair was PACOM he conducted a series of exercises with an experimentation flavor to develop the Joint Mission Force to identify and overcome barriers to Joint operations. These were excellent successes that overcame a variety of barriers (technical and cultural) to interoperability. Similarly, CENTCOM has, under the pressure arising from operations in Iraq and Afghanistan, become a popular transition target for innovation. The success of JSTARS, which was introduced in Desert Storm as an R&D system but quickly proved its worth in the field and became recognized as a useful capability, has inspired a host of efforts to place new and promising capabilities in the theater for application, assessment, and improvement. The combatant command in Korea has become a favored venue for those innovations that are expected to assist most in the type of intense combat expected there if a war occurs. NORTHCOM is seen as the desirable place to test innovations important to Homeland Defense and Homeland Security. Both TRANSCOM and STRATCOM are also seen as experimentation venues for innovations related to their missions. However, the focus of the Combatant and Functional Commands tends to be relatively short-term because their primary missions are quite immediate. Hence, they are difficult places to organize and execute campaigns of experimentation focused on transformational and disruptive innovations.

The Joint Forces Command (JFCOM) was chartered (15 May, 1998) by the Secretary of Defense as the Executive Agent for Joint Experimentation. This charter explicitly recognized that DoD experimentation needed to focus on the full range of elements required for successful mission capability packages[109] rather than the historical approach that focused primarily on platforms and technologies. This new responsibility was a logical extension of the existing JFCOM mandate to serve as the Joint Force integrator and its role in training for Joint operations. JFCOM has spent the intervening years in both

---

[109] DOTMLPF: doctrine, organization, training, materiel, leadership, personnel, and facilities.

developing its organization and the processes necessary for successful campaigns of experimentation. It has conducted a wide range of experimentation from major venues such as Unified Vision 01 and Millennium Challenge 02 to much smaller Limited Objective Experiments focused on specific issues, coalition efforts in the Multinational Experiment series, and cooperative experimentation venues with the Services such as Unified Quest and Unified Endeavor. They have been responsible for the support and extension of the Combined Federated Battle Laboratory (CFBL) that enables broad virtual experimentation with allies and for innovations such as the Joint Inter-Agency Coordination Group (JIACG) in operational-level experimentation. They will continue to be a major focus for experimentation, particularly because the JFCOM Commander is now dual-hatted as the Commander, Allied Transformation Command (ACT), which is taking the lead on NATO transformation and experimentation.

MISADVENTURES IN CAMPAIGNS OF EXPERIMENTATION

One of the most important chapters in the original *Code of Best Practice for Experimentation* was entitled, "Adventures in Experimentation." That essay identified and explored "the types of problems experienced within the past several years." However, because the purpose of that chapter was "to inform and teach, no individuals, teams, organizations, or agencies" were identified by name and some effort was made to write at a level of abstraction that would mask the specific experiments from which the materials were drawn. Moreover, to avoid focusing on unusual circumstances, every type of mistake described had to have been observed at least twice within DoD. That material has apparently served its purpose well. Any number of people and teams have reported that they found the material clear and were able to relate it to their own work. As a result, we have included similar material here, but under the somewhat less facetious title of "misadventures." This discussion, however, is much shorter than the one in the earlier book both because there are many fewer campaigns of experimentation than there are individual experiments and because the problems seen in many campaigns are often really difficulties in the individual experiments. Hence, those responsible for the overall management of campaigns would be wise to review that material carefully.

One important misadventure mentioned in the earlier list is allowing campaigns of experimentation to be governed by unrealistic schedules. The "tyranny of the calendar" usually occurs for very understandable reasons such as scheduling resources efficiently and seeking to move ahead rapidly. It is closely related to another major problem common in campaigns: *a failure to recognize that unexpected results, even evidence clearly showing that an innovation does not have the expected positive impact (in other words, a substantive failure) are not bad things in a campaign.* Such findings are all but unavoidable in efforts to identify and assess truly disruptive innovations. As such, they are valuable contributions to a better understanding of the phenomenon under study. These "negative outcomes" sometimes highlight limiting conditions (circumstances where the innovation will not work or must be modified in order to work), sometimes they identify poor or inappropriate instantiations of otherwise useful concepts, and sometimes they simply identify a concept that does not work as expected and should not be pursued further. Obviously, the capability to manage campaigns in ways that take full advantage of this substantive learning is an essential skill in working toward successful transformation. This includes that ability to manage the schedule when unexpected results (positive or negative) are generated, and not be managed *by* the schedule.

Perhaps the most common error in campaigns of experimentation is the failure to create a broad or deep enough peer review panel or to use it well. Problems in this area include forming the peer review group from the senior researchers responsible for the experiments (which means they are asked to do objective review of their own best efforts or placed in a position where they feel they should demonstrate their work is better than that of the other research teams in order to guarantee positive budgetary consideration), forming it primarily from senior officers or senior retired officers (which means the science and experimentation expertise are not involved), or failing to formulate it at all. When peer review groups are formed, they often become involved too late in the process. Their great values lie in (1) helping to craft the campaign plan in the first place and (2) asking potentially embarrassing questions early enough that they can be answered well and responsibly before the results are final. Delaying their involvement until late in the process reduces their ability to play these important roles.

Use of a single scenario or class of situation across a campaign of experimentation remains commonplace within the Department of Defense and is a serious error. This choice is often justified on the basis of cost, however, it results in reversion to threat-based analysis (rather than capability-based analyses) and it means that the innovations under study become optimized for a single situation rather than assessed in the range of contexts in which it must later be applied. Even a single experiment should always make some effort to sample the interesting space, but for a campaign not to do so is foolish.

A problem common to the Department of Defense campaigns that is largely avoided by academic and industrial experimentation programs is failure to publish the reports, artifacts, and data for review and use by other teams. Building cumulative knowledge requires sharing it so that others can build on it. Campaigns of experimentation also generate valuable artifacts such as scenarios, measurement instruments, and ways of integrating data to highlight important issues. These can also be either reused or built upon by other campaigns, saving the community creative energy, time, and money. Data collected for one campaign can often also be valuable to others either as a baseline for comparison or to highlight issues that need further research. In some cases, cross-experiment analysis will also use new insights. However, the data generated in campaigns run by elements of DoD are almost never made available to the larger community, even those researchers who have clearances and the capability to handle classified information. This is an expensive (wasteful) practice and also minimizes the knowledge gained in each campaign.

Finally, failure to properly resource campaigns remains a common problem, both in terms of total resource availability and in terms of the flexibility with which funding is administered. The current DoD budget system for research and development assumes that the processes of innovation and experimentation will proceed predictably. Hence, when an early experiment in a campaign identifies an opportunity for early or immediate exploitation or generates findings that indicate a need for either confirmation experimentation or looking at issues again (negative findings), the obviously needed adjustments cannot be made in a timely manner. Campaign financing needs to move from the current detailed planning to something approximating "block grants" with the power to reallocate funds passed on to the senior official responsible for the campaign. In addition, more funds need to be available to support rapid transition of powerful innovations into the hands of those who will use them in the field.

BARRIERS TO TRANSFORMATIONAL CAMPAIGNS OF EXPERIMENTATION

A review of DoD's experiences in developing and executing campaigns of transformational experimentation reveals several systemic barriers to success.

First, the process of innovation and capability development has been decomposed (in a traditional Industrial Age way) into discrete, sequential steps. Whether this is based on the old twelve steps for procurement based on the level of maturity, the seven more recently developed R&D Categories recognized in the DoD budgetary program codes, or the simple bifurcation used by JFCOM (Concept and Prototype), this decomposition creates institutional barriers to capability development. While a number of discrete steps can form useful milestones during a campaign of experimentation, the way phase transitions have been implemented in DoD serve primarily as budget gates that must be negotiated and not as part of a coherent process of experimentation. As a consequence, current DoD campaigns of experimentation are organized for short-term rather than long-term success. Furthermore, they are frequently interrupted at these "break points."

Equally perniciously, the personnel associated with an innovation and organizational responsibility for it also tend to change as each milestone is achieved. Consequently, the developed expertise is often lost, and with it, crucial momentum.

Momentum is crucial for overcoming the barriers to change in a conservative set of institutions like the DoD and the U.S. Government. Hence, continuity of key personnel is a very important element in planning successful innovation.

Third, the lack of serious commitment to objective measurement and hypothesis testing that is manifest within DoD makes it extremely difficult to build the evidence necessary to pursue a campaign of experimentation, convince skeptics, and ensure funding continuity. The very arguments most often used to avoid serious testing—that warfare, conflict avoidance, and conflict management are messy and difficult arenas—cry out for the development of innovations that are strong enough both (1) to matter when important decisions must be made under stressful conditions and (2) for investing the funds necessary to observe and measure what is happening when proposed innovations are implemented in experimentation settings. Results need to be first accurate and reliable and second credible to those responsible for implementing them. While credibility is currently measured using subject matter experts and survey tools, this is not the best approach to establish the accuracy and reliability of experimentation results.[110] The use of SMEs and survey tools has serious limitations, particularly when disruptive innovations are involved, and these approaches cannot be allowed to crowd out the measurement approaches that are better suited to establishing what actually happened and what has been learned.

Our emphasis on objective empirical evidence in the context of a conceptual model should not be understood as a call for applying the Industrial Age concept of formal Test and Evalu-

---

[110] Alberts et al., *Code of Best Practice for Experimentation.* pp. 161ff.
NATO Code of Best Practice for C2 Assessment. pp. 99-100.

ation processes to campaigns of experimentation. Those processes are structured to determine whether specific (pre-established) sets of requirements have been met, and as currently practiced fail to examine the agility of the innovations under development. However, the standards of measurement and types of statistical analyses used in some Test and Evaluation contexts are very useful and should be used (1) to assemble the evidence required and (2) to choose among alternative approaches.

State of the practice summary

As we have shown with the examples from DARPA and ONR, campaigns of experimentation have long been a tradition in the research and development organizations of the Department of Defense. However, those campaigns have typically encountered barriers impeding the transitions of their innovations into the force. These barriers have been both organizational and budgetary.

Over the past decade, the need for transformational campaigns of experimentation has become increasingly obvious. Network Centric Operations, Effects Based Operations, and more technical capabilities (e.g., precision munitions and UAVs), as well as the increasing need to deal with a variety of missions involving interagency, coalition, and non-governmental actors create demands that cannot be satisfied without fundamental changes in the way we do business. As the cases of the Army's Stryker Brigade and Blue Force Tracker illustrate, we are deploying new capabilities almost as rapidly as we can develop them. Moreover, the development of the Information Age capabilities that are both possible and badly needed in the field is still in its early stages—much more capability can and will be generated. Hence, we cannot afford the breaks in momentum that come from an inability to develop *and* transition mission capability packages as part of a coherent, continuous campaign of experimentation.

Campaigns of transformational experimentation therefore need to be planned for success, with research and development programs that are:

1. based on the rigorous application of the basic principles of science, measurement, and experimentation;

2. based on intelligent model-experiment-model approaches;

3. agile enough to adapt as new learning takes place;

4. disseminated broadly for peer review and multiple impacts; and

5. resourced adequately for success.

CHAPTER 8

HARVESTING RESULTS

The challenges of transformation are daunting because of their scale and complexity. The scale involved in transformation is formidable because transformation requires that virtually every aspect of "who we are" and "what we do" be examined through a new lens (a combination of network-centric organizations and approaches and new mission challenges). Thus, transformation reaches into every organizational element of DoD. Given the nature of an Information Age transformation, the network-centric mission capability packages developed to accomplish new tasks or new ways of accomplishing old tasks need to coevolve technology with concepts, organizations, approaches to command and control, education, training, and the rest of the elements of MCPs. Thus, not only does transformation involve every DoD organization, but, at a minimum, requires collaboration among significant numbers of DoD organizations and will, in many cases, also involve collaboration with non-DoD organizations, other USG offices, coalition partners, international organizations, NGOs, and PVOs. Hence, transformation requires a rich set of interactions among traditionally stovepiped organizations. This is a major source of complexity.

Furthermore, DoD consists of more than the sum of its mission capability packages. In addition to the resources that are devoted to MCPs, the enterprise requires a set of enabling processes and capabilities that (1) recognize the need for particular mission capability packages, (2) budget, plan, and program for the resources necessary to create and maintain these MCPs, (3) develop and acquire the capabilities needed, (4) recruit and retain personnel, and (5) educate and train. These business processes also need to coevolve along with the mission capability packages they support. This adds an additional layer of complexity to the task.

What do these scale and complexity challenges mean for campaigns of experimentation? The scale of transformation translates into the number of campaigns of experimentation that are needed and the complexity translates into the need to harvest results across these campaigns.

CAMPAIGN SYNERGIES

Successful campaigns of experimentation result in new capabilities that are embodied in mission capability packages or add to our collective knowledge, a prerequisite to developing new capabilities. The success of an individual campaign ultimately depends upon both its ability to absorb and build upon the existing body of knowledge (experience) and its ability to focus experimental activities (experiments and analyses) to fill gaps in understanding (key variables, relationships, and conditions) or to develop new approaches. Successful campaigns need to adjust their plans continually as the results of experiments and analyses become available.

In addition to being able to harvest experimental results within the context of a single campaign, these results need to be more widely understood and utilized. Experimental results need to be shared to (1) cross-fertilize mission capability focused campaigns,[111] (2) build a coherent body of knowledge (experience) related to the nature of an Information Age transformation, and (3) guide the transformation of enterprise functions and services that directly and indirectly support the development and employment of mission capability packages. Thus, successful campaigns are necessary but not sufficient to achieve transformation because individual campaigns need to be augmented by cross-cutting efforts that are designed to share and harvest results.

As is the case with individual experiments, individual campaigns are limited in what they can accomplish. While individual experiments can only provide fragments that need to be combined with other fragments, individual campaigns provide pieces that need to be assembled into larger constructs. Experimentation campaigns provide the mechanisms to combine fragments into pieces. They can be thought of as programs that deliver individual systems that need to be integrated into a federation of systems.[112] The relationship between and among systems in a federation cannot be wholly prescribed in advance but must evolve over time. We must anticipate that many transactions (or threads through individ-

---

[111] We anticipate that DoD, together with its counterparts in government, coalition partners, and others will need to collaboratively conduct a large number of campaigns of experimentation. This book briefly discusses the nature of the campaigns that will be needed in the future, but a detailed treatment of the campaigns that could or should be pursued has been left for another time.

[112] The term *system of systems* is often used, but this term implies a much more closely coupled set of systems than is desirable in a network-centric world.

Campaign synergies

ual systems) cannot be predicted and hence we cannot specify these in advance nor develop tests that can adequately stress individual pieces (systems or tactics), much less be able to define or predict a constantly evolving federation. In a similar vein, we cannot predict how individual campaigns will unfold other than to know that they will generate knowledge and develop new capabilities. The knowledge and capabilities generated will create the opportunities for synergies that can be realized across campaigns. Thus we know we will get answers, but we do not know what the answers will be or the questions that will subsequently be raised.

The potential synergies across campaigns can be thought of as having the same properties as the interactions among systems in a federation of systems. That is, we can identify many or even most of the interactions that currently take place but we cannot predict with any degree of accuracy what interactions might take place as the systems that comprise the federation change, evolve, and are used. Therefore, if we focus only on what we can specify in advance we will be missing a significant number of opportunities—perhaps the most important ones.

The Power to the Edge vision recognizes this reality and moves away from an approach to interoperability based upon making applications (systems) interoperable with one another by specifying information exchange requirements to an approach that is based on data interoperability. The corresponding governance model is based upon the development of community of interest data tagging and related standards that promote data reuse. The net effect is that data become the currency of the federation of systems. The data produced by one system or application previously "tunneled" to one or a small number of associated applications is instead posted. This makes it available to whatever application is aware of the availability of these data or has the ability to discover them, provided that access is granted. Thus, the economic barriers that have existed for so long to the development and adoption of new applications or applications that were tailored to specific users are greatly reduced. This is because new applications create no burden on existing applications, nor are they required to develop application-specific interfaces for all of the systems that provide inputs to them or those for which they may provide inputs.

An individual campaign of experimentation is analogous to a system that can be managed because the experiments and analyses that are conducted as part of a campaign can be focused, sequenced, and the data elements standardized by a program manager who can make trade-offs in the pursuit of a given capability or element of knowledge.

Creating the conditions favorable for harvesting experimental results across a diverse set of campaigns is, analogously, a governance issue. At the risk of over-simplification, the results of experiments and analyses are, in essence, data. The same set of principles, policies, and governance regime that creates the conditions for wide-spread information sharing and collaboration within DoD and between DoD and other organizations should work to foster an environment that facilitates and encourages the reuse of empirical data and findings across campaigns of experimentation.

Cross-campaign synergies can either be direct or indirect. Direct synergies are most likely to occur among campaigns of the same type—those that are conducted within a community of interest. DoD campaigns of experimentation can be coarsely grouped into the following three types or communities of interest, each of which will consist of many overlapping sub-communities:

- Campaigns that seek to apply the tenets of NCW to specific military operations;

- Campaigns that seek to apply network-centric principles to DoD business processes; and

- Campaigns that seek to build a body of knowledge related to concepts embodied in Information Age transformation, e.g. the nature of the relationships among variables.

Campaigns of the first type are primarily conducted by military organizations that focus on the development of concepts, doctrine, and operations. Campaigns of the second type are primarily conducted under the banner or DoD business process re-engineering. Campaigns of the third type are undertaken by DoD and DoD-sponsored organizations that focus on research and development. There is an increased likelihood of direct synergies within each type because whatever cultural barriers may exist within each of these communities is considerably less than the cultural and linguistic differences among them.

Within the community that seeks to apply NCW (NCO, NEC, EBO) to military operations, cultural differences exist among the various Services, between Service perspectives and the Joint community, and among allies and coalition partners. However, these differences are outweighed by a shared understanding of the nature of combat, mission outcomes, and warfighting experiences. This should allow these types of campaigns to directly use the results obtained in one another. These mission capability concept-based experiments have been initially focused on similar questions. For example, at our current level of NCW maturity, many experimentation activities are focused on understanding the implications for doctrine and organization of widespread availability of near-real-time information and improved shared awareness.

Our efforts to date have been hampered by shortfalls in the number and nature of the campaigns being undertaken and a lack of mechanisms that support sharing of experimental data and results. Considering the nature of an Information Age transformation of DoD, (1) there are far too few experiments being conducted, too many of which are not part of coherent campaigns of experimentation, (2) there are far too few campaigns being conducted, and (3) there is a seeming reluctance to explore concepts, particularly command and control, outside of a small comfort zone.

The situation is only marginally better with respect to the R&D community where the current focus is on a relatively short list of research objectives: being able to measure key concepts, developing an overall conceptual framework, characterizing the functions associated with command and control, and testing hypothesized relationships that are embodied in the tenets of NCW. As in the MCP-focused community, there are far too few experiments being undertaken, too few of which are part of campaigns, and some lack imagination. But in the research community there have been two positive developments that will facilitate synergies among experimentation activities. If adopted by the concept development community, they have the potential to improve the ability to generate synergies among MCP-focused campaigns of experimentation as well. These developments involve collaborations between the OASD(NII) (Office of the

Assistant Secretary of Defense for Networks and Information Integration) and OFT (Office of Force Transformation) on a network-centric conceptual framework and between NATO and OASD(NII) on explorations of network-centric approaches to command and control. As a result of these collaborations, a language for understanding network-centric principles and operations is emerging along with a set of metrics that can be used to characterize new concepts, approaches, and organizations as well as measure the degree to which the principles or tenets of network centricity are being realized and how this relates to success. This common language and set of metrics puts the international research community in an excellent position to develop synergies among campaigns of experimentation being undertaken around the globe. They also satisfy a basic requirement for building a body of knowledge.

BUILDING A BODY OF KNOWLEDGE

While direct synergies among campaigns are certainly important, most of the synergies that will occur will be indirect. Indirect synergies occur when the data collected or lessons learned from campaigns of experimentation are first abstracted, synthesized, and incorporated into a conceptual framework, model, or archive that reflects the current state of the art or practice. Then the model or lessons learned archive is used as a point of departure for a given campaign or individual experimental event. For example, the relationship between information sharing and shared awareness under specific conditions has been determined by analytical synthesis of empirical evidence collected from a number of experiments and incorporated into a model or conceptual framework. This knowledge can then be utilized by campaigns in various ways.

While the aforementioned collaborative efforts involving OASD(NII), OFT, and NATO and related efforts that involve researchers around the world are making progress toward a unified conceptual framework that will maximize synergies among research efforts, there are simply not enough resources or efforts being devoted to these kinds of activities to keep up with the data collected and experiences accumulated. Efforts to develop a widely accessible experience archive, if they are being undertaken, are not known to us.

While a good foundation for building a body of knowledge has been laid, progress will depend on the extent to which empirical data are collected, shared, analyzed, and archived. Resources will need to be reallocated to reach a better balance among individual experiments, campaigns of experimentation, and activities devoted to building bodies of knowledge and experience. At this point, resources need to shift from a primary focus on the conduct of individual experiments to the conduct of campaigns of experimentation. In addition, relatively modest investments need to be made in the development and maintenance of conceptual frameworks populated with empirical data and analysis results and retrievable archives of experiences.

CHAPTER 9

THE WAY AHEAD

W hile experimentation activities are taking place throughout the DoD, these are, for the most part, individual
events that have not been conceived or executed as parts of
coherent experimentation campaigns. As a result, most of these
activities have not been able to build effectively upon one
another, nor have they collectively been adequately able to
explore the complex issues that we face today in the twenty-first
century, nor have they provided adequate assurances that new
concepts will perform as expected.

The recognition of the need for campaigns of experimentation
is growing[113] and organizations like JFCOM are planning and
executing concept-based campaigns. But these activities, as
currently conceived and carried out, are insufficient to support
transformation. To support transformation adequately, current
experimentation activities—both the campaigns of experimentation that are underway and planned as well as the individual
experimentation activities they are built around—will need to

---

[113] As indicated in the Acknowledgments, it was General Dubik, the J9 at
JFCOM, who requested that we undertake the writing of this book.

be improved, better connected, have a broader focus, and be conducted in greater numbers.

BETTER EXPERIMENTS

Individual experiments are the building blocks of campaigns of experimentation. While significant progress has been made by DoD organizations toward improving the design, planning, and conduct of individual experiments since the publication of the *Code of Best Practice for Experimentation,* there is still a great deal of room for improvement. New material for the chapter "Adventures in Experimentation" in that volume continues to be generated by experimentation activities despite being acquainted with the "adventures"[114] of others found in the *COBP for Experimentation.*

Experimentation often continues to be confused or confounded with training exercises. The basic incompatibility between the two is not fully understood. Training, as it is traditionally practiced, assumes one knows the right or best way to accomplish a set of essential tasks and ensures or reinforces knowledge about that current practice or doctrine. Experimentation assumes that a good way to accomplish something needs to be discovered or developed (and that it will probably evolve over time). Given the widely expressed need for transformation, it follows that we do not know the best way to accomplish the tasks at hand and, in some cases, may not even know what the specific tasks are that need to be accomplished. This does not mean that there is no room or role for training during a period of transformation, but rather that care must be taken to balance the two different kinds of activities and to

---

[114] Perhaps better understood as *misadventures.*

keep an open mind with respect to how traditional tasks should be approached. Currently, we have not achieved an appropriate balance.

It takes more than a few knowledgeable individuals in an organization to conduct quality experimentation activities—and it takes more than good intentions. The ability of an organization to conceive, design, plan, and conduct quality experiments and then fully exploit the data they yield requires understanding, commitment, training, and experience. This is even truer of campaigns of transformational experimentation. Presently, few DoD organizations have sufficient levels of these essential attributes. To remedy this situation, policies and programs are needed to:

1. Develop a culture of experimentation.

2. Establish experimentation as a core DoD competency.

3. Ensure that adequate resources are available.

4. Increase understanding about the role of experimentation in transformation.

5. Develop knowledgeable individuals at all levels to design, plan, and conduct experiments.

6. Establish standards of excellence.

7. Encourage the sharing of the data collected, the analyses undertaken, and the lessons learned regarding experimentation.

ESTABLISHING THE PREREQUISITES FOR SUCCESS

The achievement of a scientific breakthrough, the associated development of a new theory, and the application of such a theory that results in new practices and products require both assembling and integrating many facts across a large number of steps. Therefore, the speed at which these advances are achieved depends on the pace, focus, and connectedness of individual efforts.

Discovering and piecing together the relevant facts, drawing conclusions, and making inferences are somewhat analogous to trying to assemble a huge jigsaw puzzle for which many of the pieces do not fit exactly and for which many pieces have yet to be created. In the case of DoD's experimentally based transformation, we are trying to assemble a significant number of puzzles. Some are concept-based, as in the case of trying to instantiate the theory of Network Centric Warfare or Effects Based Operations. Some are knowledge-based, as in the case of trying to understand the driving factors that enable successful collaboration or situation awareness.

While the kind of institutionalized campaigns of experimentation that we are advocating for the purpose of quickening the pace, improving focus, and increasing connectedness is a new idea for the DoD, experimentation activities have been orchestrated, self-synchronized, or at least connected for almost as long as experiments have been conducted. There exists a cultural value in science that virtually mandates that experiment results are to be published and made accessible to other researchers (and increasingly, shared with the general public). The rates of scientific progress we have experienced would not have been possible without a culture of sharing, peer review, recognizing individual achievements, and building upon the work of others.

Truly original work, which draws upon no other previous work, is extremely rare. There are two major reasons to share information and insights and to work collaboratively. They both have to do with wealth—the fact that no one person (or organization) has enough wealth, intellectually (in knowledge and perspectives) or materially (in resources and time), to accomplish the tasks at hand if everyone started from scratch and pursued their investigations independently.

Throughout most of history, sharing information about experiments has been very limited. First, there were not very many individuals who could properly understand the significance of experiment results, and second, the means of communication were rudimentary. Thus, information sharing regarding experiments and their results was generally limited to the few individuals known by the experimenter to be knowledgeable and interested.[115] There were also few institutions that were both wealthy enough and interested enough to fund a critical mass of research for an indefinite period of time except in a few areas, such as military technologies.[116]

Today, results are widely available so quickly and so widely that there is increasing concern that more time should be taken for review.[117] The number of knowledgeable individuals has grown exponentially with increases in literacy, advanced degrees, and population growth. We are familiar with the tremendous rate of scientific advances in the last century. The rate of these advances has been linked to these fundamental

---

[115] The limits of a *smart smart push* approach to dissemination of information are discussed in: Alberts and Hayes, *Power to the Edge*. pp. xiv-xv, 75-77.

[116] Note that it has been traditionally technologies that have received funding, not research into the other aspects of mission capability packages (e.g., approaches to command and control or concepts for peace operations).

[117] Posting in parallel is the current OASD(NII) policy, and on balance with proper metadata tagging is a better approach than attempting to control information dissemination. Alberts and Hayes, *Power to the Edge*. p. 82.

trends, to our improving ability to disseminate results, and to the increasing number of knowledgeable people available to understand and utilize them.

In selected domains, campaigns of experimentation are largely self-synchronizing because there are well-established disciplines with well-qualified members that have articulated priority research questions and issues, and because mechanisms are in place to promote information sharing and collaboration. In areas where there are no long-established disciplines (generally offshoots of existing disciplines and interdisciplinary domains), researchers have been recently helped by the evolution of the Internet. The Web provides individuals with greatly improved means for finding material of interest, as well as a quick and low-cost means of information dissemination. This enables the formation of new associations of interested individuals and the coalescence of communities of interest.

Established scientific disciplines and research institutions are very good at articulating the puzzle that they are trying to solve, as well as discovering and cataloguing available pieces, and focusing their efforts on fashioning needed pieces. In the case of DoD, an understanding of what pieces of the puzzle currently exist will depend upon the efforts made by the organizations involved in transformation to share the results of their experiments and the efforts made to research related experiences in other domains (e.g., business). Currently, these efforts are inadequate for the task at hand. Too many "reasons" are found to delay (and often to prevent) broad dissemination of results. This reflects the clash of two traditions: military culture, in which only success (narrowly defined) is acceptable, and scientific culture, in which "failure" provides important information.

Once one knows what the puzzle is and what pieces are currently available, attention turns to creating the missing pieces. Creating new pieces is a function of the experimentation activities that are undertaken. The degree to which these activities create pieces that are relevant and that fit well into one or more puzzles depends upon how well they were conceived, as well as how well they were designed and executed. The better connected experimentation activities are (i.e., the more information about them that is shared), the more relevant and useful their results (the pieces they create) will be.

Connectedness involves far more that information sharing. It involves the establishment of domain or community of interest goals (the puzzles that need to be solved) and mechanisms to ensure that the experimentation activities being conducted are conceived with the knowledge of what has been learned to date and what needs to be learned to make progress towards shared community goals. Thus, each experimentation activity needs to be designed and conducted with more than the immediate needs of the organization that undertakes the experiment in mind.[118] When each organization spends more on their experimentation activity (e.g., to consider another variable, to look at an extra treatment, to ensure that the experiment generates useful information, to document the results), the enterprise as a whole accomplishes more and ultimately spends less.

In the case of DoD, the conditions already exist for *connectedness* in the area of military technologies with well-established disciplines and sources of institutional funding, but these conditions do not extend to the area of transformation. Therefore, the

---

[118] The current practice of piggybacking continues to thwart progress. Alberts et al., *Code of Best Practice for Experimentation*. p. 56.

development of a comprehensive set of campaigns of experimentation, with the involvement of key DoD organizations, is necessary to achieve the degree of connectedness that has been achieved in a self-organizing manner in other areas and domains. One of the primary reasons for this is that while the future of the military is generally agreed to be Joint, Joint experimentation and related research is in its infancy and coalition-related efforts barely exist. This is exacerbated by the fact that the vast majority of research and experimentation on new concepts that is being conducted by the Services and other DoD organizations appears to be generally focused inward. Jointness is more than the sum of its parts. Inwardly focused experimentation will not create the right pieces for the puzzles of transformation. Thus, while efforts have been made to increase information sharing and connectedness, far more needs to be done in this regard if DoD is to transform.

CAMPAIGNS OF EXPERIMENTATION

Campaigns define a puzzle or set of puzzles to be addressed and the activities to be undertaken to identify existing pieces, create the missing pieces, and assemble the pieces. The nature of the campaigns that DoD needs to undertake to transform itself into an Information Age organization have yet to be adequately articulated. Adequate funding for the development of organizational competencies to carry out these campaigns and orchestrating these campaigns across DoD also need to be addressed. Of these two major issues that affect our ability to undertake transformational campaigns, the nature of the campaigns that need to be undertaken must take precedence. If we make progress here, then at least we will be looking in the right places and addressing the right problems: a set of interrelated issues that individually and/or in combination should guide the formulation of the campaigns of experiments that will be conducted by DoD in the coming years.

CHALLENGES TO ORGANIZE CAMPAIGNS AROUND

The challenges discussed in this section, while not an exhaustive collection, are among the most important subjects that need to be systematically investigated in order to fashion new network-centric mission capability packages and contribute to the emerging body of knowledge related to Information Age concepts. Creating new network-centric MCPs and building a body of knowledge regarding Information Age concepts and their applications are synergistic activities. New MCPs provide proofs of concept and opportunities to collect empirical evidence. Building a body of knowledge allows us and others to create better MCPs and accelerate the ongoing Information Age transformation taking place in organizations around the world in many different domains. The sections that follow discuss challenges related to the elements of MCPs that need to be coevolved in the context of specific mission characteristics to meet the range of missions that might lie ahead. Specifically, these are challenges related to new concepts of operation, organizational forms, and approaches to command and control that will allow us to leverage the power of information.

THE CHALLENGE OF MISSION DIVERSITY

The variety of missions that DoD is and will continue to be called upon to undertake is, in and of itself, one of the greatest challenges of the twenty-first century.[119] For some of these mis-

---

[119] Planners should consider missions that are "improbable but vital" as well as unlikely missions precisely because we are prepared to undertake them.

sions, the military will have the lead, while for others the military may be in a supporting role. Some of these missions will be accomplished with predominantly U.S. military forces; many will require a military coalition, while others will require close cooperation among civil and military forces, as well as non-governmental and international organizations.[120]

Thus, these missions differ significantly in both the nature of the tasks that need to be undertaken, the number and diversity of the participants, and the constraints that exist. These differences have a profound effect on the nature of the information and the information flows required to develop shared awareness and the relationship between shared awareness and effectiveness.

For example, because traditional combat missions differ significantly from stabilization operations, a particular organizational form or approach to command and control that may be suitable for one may not be suitable for the other. In reality, different kinds of operations will occur simultaneously or overlap in time. They are often in close proximity to one another and may involve some of the same forces elements.

THE CHALLENGE OF AGILE ORGANIZATIONS

It would be highly desirable to find organizational forms and approaches to command and control that can either perform well across a range of mission types or can dynamically adapt to changing situations by adjusting information flows, processes, and delegations of responsibility. We have learned from a variety of operations that traditional military organizations,

---

[120] Alberts and Hayes, *Power to the Edge.* pp. 107-120.

Challenges to organize campaigns around

command and control, and doctrine do not work well in some situations. We have also learned that we cannot limit the use of military forces to situations that are purely military with coalition partners that are willing to do things our way. These operations offer us valuable insights and opportunities to try to modify the way we do things, but "experimenting" within the context of these operations will not enable us to fully explore the opportunities that Information Age concepts and technologies offer. Not only are these opportunities limited in their number, but also they are limited because of the risks involved in pushing innovations to their breaking points.

THE CHALLENGE OF ACHIEVING SHARED AWARENESS

Moving to a network-centric force is conceptually simple, but "the devil is in the details."[121] The tenets of NCW can be viewed, in simplistic terms, as a two-step process. First, one creates the conditions necessary to achieve a high level of shared awareness, and then one creates the conditions to leverage shared awareness by moving power to the edge. Creating shared awareness and leveraging it are both challenging.

The tenets of NCW hypothesize that a robustly networked force will lead to improved information sharing and that this improved information sharing and collaboration will result in improvements in both the quality of information and shared awareness.[122]

Applying these concepts to civil-military missions in a coalition environment means that the nature of the force is significantly different from the force that may be required for

---

[121] Common adaptation of Mies van der Rohe's: "God is in the details."

[122] Alberts, *Information Age Transformation*. pp. 7-8.

combat operations. This means that the "denominator"[123] in all calculations involving information sources and sharing, collaboration, and shared awareness is larger and more diverse. Understanding the impacts of scale and diversity is therefore a key issue.

The degree of diversity that exists has a potential impact on a number of enabling conditions for information sharing, collaboration, and the ability to develop shared awareness. Diversity brings with it a decreased likelihood that participants will view (perceive) things the same way, trust each other, share the same set of experiences and hence interpret things the same way, share values, and share norms of behavior. Lest one think that diversity is a problem to be avoided, diversity (despite the obvious challenges that it involves) also has some very positive attributes. Diversity brings more perspectives to bear and is less likely to miss some key information, more likely to consider a wide range of options, at less risk of *groupthink*, and less likely to settle prematurely on a course of action. Diversity is particularly important when dealing with complex problems for which the courses of action must be developed from scratch, and the criteria for choosing among them are also driven by a creative process. Diversity also increases the chances of being able to understand and deal with diversity itself (a positive feedback) and more likely to understand an asymmetric adversary.

Sharing information widely, particularly the *post and smart pull* paradigm that is a central part of DoD's Power to the Edge vision,[124] is counter-cultural to many individuals and organi-

---

[123] The number of actions involved. See the metrics in: Office of Force Transformation. "NCO Conceptual Framework Version 1.0." pp. 17-52.

[124] "Transforming America's Defense." Pamphlet, OASD(NII). 2004.

Challenges to organize campaigns around

zations. Policies and directives alone will not change behavior thoroughly enough or quickly enough to make this vision a reality. Understanding how to change individual and organizational behavior by a combination of incentives, education, training, and well-presented evidence will be crucial.

Trust is generally thought to be a necessary prerequisite for information sharing. How trust can be built and transferred from individual to individual and from organization to organization is a key question. Establishing trust in a networked environment, when individuals do not know each other and interact only in cyberspace, will be critical. At the same time, we must be able to avoid inappropriate trust. One of the enablers of trust is confidence in the identity of the individual/organization and the security of the information and the transaction. Thus, security classification and information assurance policies, processes, techniques, and tools will, to a large extent, determine how much information can be shared and will be shared. One of the reasons that it is difficult to reconcile the desire to protect information and to share information is the lack of real evidence as to the nature of the expected compromises that accompany different approaches and their consequences.

We know we must move from an approach to security that is based on the minimization of risk to a more balanced risk management approach, but we do not have the empirical evidence to support policy analysis. We know that we need to move from a very gross classification system that focuses on clearing individuals to an approach that is more fine-grained and is focused on transactions. But we have neither the ideal technology nor sufficient experience with emerging technologies to give us any comfort that we can do this well enough. If

Challenges to organize campaigns around

we wait until either a perfect technology is developed and tested or until we have a definitive analysis of the consequences of increasing information sharing using currently available technologies, we are forcing ourselves to forego many of the benefits of information sharing. The more we understand just how valuable information sharing is and could be, the more likely it is that we will relax overly restrictive policies and practices. Thus, establishing the value chain from improved information sharing to improved information quality, awareness, and shared awareness should be a high priority.

THE CHALLENGE OF LEVERAGING AWARENESS

It is axiomatic that, all other things being equal, better information will result in better task performance. Individual decisionmakers throughout an organization who have improved access and/or better information are more likely to make better decisions. The value of better information is clearly situation- and scenario-dependent. A better understanding of the contribution of information to mission effectiveness would serve to provide a quantitative basis for decisions related to the provision of an Information Age infostructure (e.g., GIG-related investments). While the gains associated with this improvement in information quality may be quite large (e.g., less shots per kill can make the difference between mission success or mission failure and also ensures greater residual capability and smaller logistical demands) these are, nevertheless, incremental improvements and they pale in comparison to what is possible when shared awareness can be leveraged by new network-centric concepts of operation and new Power to the Edge approaches to command and control (changing the way we do business).

Challenges to organize campaigns around

The issue of how improved information contributes to decisions and to mission effectiveness has been studied for quite some time (e.g., sensor to shooter studies). On the other hand, the value propositions related to network-centric operations have not been as thoroughly analyzed. This is because there is only limited experience with network-centric concepts and their applications. Thus while we have a growing body of evidence[125] that network-centric operations can result in dramatic improvements, we have little well-documented quantitative assessments. Further, this network-centric experience is limited to command and control approaches that are fairly traditional. Almost none of the experience is with true Power to the Edge command and control approaches. The simple fact is that we have barely started to explore new command and control approaches.

For example, the early experiments that JFCOM and others have been doing focus on how to reinvent a headquarters. Assuming the existence of a headquarters greatly restricts the space of exploration. Even if a headquarters, in some form, turns out to be useful, it is not clear at this point even what functions a headquarters needs to perform in a network-cen-

---

[125] See: Alberts et al., *Understanding Information Age Warfare.* pp. 239-285.
Ministry of Defense. "Network Enabled Capability." April 2004.
University of Arizona. Decision Support for U.S. Navy's Combined Task Force 50 during Enduring Freedom.
Reinforce. Multinational Operations (During IRTF (L) trial of AMF (L); Amber Fox; and ISAF 3).
RAND. Stryker Brigade Combat Team.
PA Consulting Group. Joint U.S./U.K. Combat Operations in Operation Iraqi Freedom.
SAIC. Air to Ground Operations in DCX (Phase 1), Enduring Freedom and Iraqi Freedom.
Booz Allen Hamilton. NCO Conceptual Framework: Special Operations Forces Case Study.

tric world. Thus, there is a set of questions that needs to be addressed that focuses on the differences between traditional and Power to the Edge approaches to accomplishing C2 functions including, for example, monitoring the situation, establishing and communicating intent, setting ROEs, assigning responsibilities, aligning resources, recognizing changes in the situation, and responding to these changes in a variety of ways. These questions deal with not only the ways these C2 functions can be performed but the conditions under which different approaches are appropriate. Indeed, the whole subject of echelons becomes an issue. In a recent interview, Admiral Giambastiani, dual-hatted as commander of USJF-COM and NATO's Allied Transformation Command, notes at one point that his primary responsibility is at the operational level of command, then moves immediately to the problem of coordinating fires across a force, which has traditionally been considered a tactical responsibility.[126]

One needs to start with an idea of the full spectrum of potential C2 approaches. *Command Arrangements for Peace Operations*[127] presents a spectrum of C2 approaches taken from history. Work is currently underway as part of the NATO effort[128] to establish the dimensionality of C2 approaches (e.g., the identification of the key characteristics of C2 that make one approach different from another). A key characteristic that distinguishes one C2 approach from another is the degree to which authority is centralized or decentralized. In some cases, for example coalition operations, there simply is no single authority despite formal tables of organization that imply precise command arrangements.

[126] Keeter, "Giambastiani: Change in Culture." pp. 35-40.

[127] Alberts and Hayes, *Command Arrangements.* 1995.

[128] NATO SAS-050. *Terms of Reference.* June 2003.

Challenges to organize campaigns around

In the extreme, there may be situations in which groups are formed with no one in charge—peer organizations. Peer or edge teams are successful when they have enough overlap in interests and shared awareness to enable them to accomplish tasks both effectively and efficiently.

In the final analysis, the nature of the situation (complexity, time urgency, nature of risk, payoff function) will determine which C2 approach works best (highest expected value, minimized chance of bad outcome, etc.). Exploring the full spectrum of C2 approaches under a wide range of mission contexts and identifying what works, when, and why are on the critical path to transformation.

Interagency, coalition operations, and civil-military missions deserve special attention. These operations are challenging for several reasons. We have already discussed the diversity of participants and the impacts that diversity can have on perceptions, values, and the ability to develop shared awareness. Equally important is the fact that different participants may have, in addition to somewhat different perceptions of the situation, different goals and objectives. Coalition building, a process that involves working with a set of very different perceptions, values, constraints, and objectives in order to establish common ground, is among the most challenging tasks we face. Often this involves fashioning an approach to C2 that all of the parties can accept. Attempts to do this within the traditional military hierarchy have resulted in coalition processes that have been less than ideal.[129] Given the range of C2

---

[129] See discussions of peacekeeping arrangements in:
Siegel, *Target Bosnia.* 1998.
Wentz, *Lessons from Kosovo.* 2002.
Alberts and Hayes, *Command Arrangements.* pp. 42-50.

approaches made possible by our ability to develop improved shared awareness, we now have new C2 options, ones that would be less problematical for potential coalition partners. We need to thoroughly explore these new C2 options. We need to understand, for example, how requisite levels of trust can be established and how far we can go with self-synchronization. There is no need, of course, to have the same approach to C2 employed for every participant. Exploring mixed C2 solutions would give us an opportunity to make our "coalition tent" more inclusive and bring more information, experience, and capabilities to bear.

ILLUSTRATIVE CAMPAIGNS

There are two major dimensions of the Information Age transformation of the DoD. The first involves the nature of the missions that we will be called upon to undertake, while the second is, at its heart, about new approaches to command and control that are enabled by a robustly networked force. Both of these represent significant departures from history and tradition. Both require innovative thinking outside the box, unconstrained by the status quo. Developing an understanding of the challenges associated with both of these dimensions and developing mission capability packages that instantiate these understandings will require undertaking a number of related campaigns of experimentation. This section will discuss three such campaigns. These will serve to illustrate campaigns in their formative stages and include campaign objectives, scope, outlines of a campaign plan, and the identification of a number of related experimentation activities. The first campaign deals with developing a better understanding of the nature of command and control in a networked environment while the second campaign is devoted to developing a command and control approach to stabilization operations. The third campaign description addresses command arrangements for homeland defense and homeland security.

COMMAND AND CONTROL
IN A NETWORKED ENVIRONMENT

It is axiomatic that different approaches to command and control will work best for different sets of participants engaged in different missions under different sets of circumstances. Thus, a transformed force will need to be able to organize and operate in a number of ways to be agile enough to meet twenty-first century mission challenges. Clearly it would be desirable to minimize the number of approaches to command and control that a force will employ to avoid confusion and enable it to develop high levels of competency. Thus, it is important that we understand whether or not some approaches to command and control are dominant over a wide range of sets of participants, missions, and circumstances or, if not dominant, are good enough to employ given the drawbacks associated with having to learn and employ a large number of approaches. To answer this question, we need to understand how different approaches to C2 are affected by the nature of the participants, the nature of the mission, and the nature of the conditions that prevail.

Understanding the nature of command and the nature of control in a networked environment will take considerable time and effort. Many campaigns of experimentation and the lessons from many operations will be required for us to be able to adequately explore the possibilities available and the potential of specific approaches. The campaign discussed below is a first step that builds upon what we have observed to date with regard to the concept of self-synchronization.

The central question to be addressed by this campaign is the relationship between shared awareness and the degree to which a force can self-synchronize. Central to all campaigns is a conceptual model that represents our understanding and embodies hypotheses. Figure 5 presents the top-level conceptual model that serves as a point of departure for this campaign.

Although this campaign conceptual model contains just a very small subset of the variables that are contained in the NCO Conceptual Framework[130] and the NATO C2 Conceptual Model,[131] the campaign itself will be quite challenging. During the course of the campaign, much will be learned regarding (1) how to characterize C2 approaches, (2) how to measure the concepts of shared awareness, self-synchronization, and the intervening variables, and (3) the nature of the relationship between shared awareness, C2 approach, and self-synchronization and the factors that influence this relationship. The results generated by this campaign will, among other things, shed light on just how much shared awareness may be needed, as a function of C2 approach, to anchor other campaigns that are focused on achieving shared awareness.

The formulation phase of this campaign will consist of a series of literature searches related to C2 approaches, team performance, the measurement of the variables contained in the initial version of the campaign conceptual model, and reviews of past experiments and lessons from operations. During this formulation phase, the conceptual model will be updated to

---

[130] Office of Force Transformation. "NCO Conceptual Framework Version 1.0."

[131] *NATO Code of Best Practice for C2 Assessment.* p. 36.

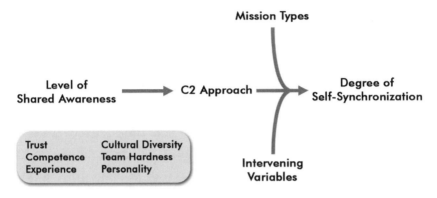

Figure 5. C2 in a Networked Environment Campaign Conceptual Model

reflect additional intervening variables, C2 approaches will be characterized in terms of a small number of variables, different C2 approaches will be identified and mapped to value ranges for these variables, and measures associated with the conceptual model variables will be operationally defined.

In the initial experimentation phase of the campaign, a series of simulation-based analyses and experiments with small teams will be undertaken. The simulations will focus on exploring a wide spectrum of C2 approaches while the small team experiments will focus on the factors that influence cooperative behaviors.

Based upon the results of these analyses and experiments, the campaign will move, in its next phase, to a series of more sharply-focused analyses and experiments that involve the establishment of a baseline and the exploration of the most promising C2 approaches—that is, those that are thought to foster a very high degree of self-synchronization as well as those that can function well with different degrees of shared awareness and/or at less stressing values for the intervening variables. The campaign will then move on to explore the sensitivity of the most promising C2 approaches to changing in initial conditions.

The next phase of the campaign will involve a series of experiments with teams with different characteristics and different degrees of shared awareness. These experiments will determine whether or not we can predict the degree of self-synchronization that will emerge. This will be followed by a series of demonstrations that will serve to educate individuals and organizations through the DoD, other USG offices, allies, and other potential coalition partners.

The final phase will be to work with others to transition those C2 approaches that were found to be promising so that more experience can be gained and evidence accumulated. This transition phase needs to be built into the campaign and may well involve the development of new instantiations for the key innovations.

COMMAND AND CONTROL FOR STABILIZATION OPERATIONS

Stabilizations operations include all of the aspects of the "three block war" in which U.S. and coalition forces are simultaneously (1) engaged in ensuring security and supporting those providing reconstruction and humanitarian assistance, (2) acting as peacekeepers (and often peace makers), and (3) engaged in selected military operations against forces taking direct military actions (these may be remnants of armed forces, insurgents, or terrorists). Recent relevant experience includes East Timor, Kosovo, Bosnia, Iraq, and Afghanistan.

Command and control is a challenge during stabilization operations because they are inherently complicated. Many different actors are involved. These are effects-based operations in which the results depend not only on concrete actions, but also on people's (local, national, and international) perceptions of those actions, and the relevant command arrangements involve very different agendas and structures of authority. Key issues include the creation of "unity of intent," the synchronization of plans and actions, and the development of synergy.

The initial phase of this effort—formulation—will involve searching for previous writing and evidence on the topic. This includes existing doctrinal publications (both the U.S. Army and USMC have relevant materials, as do the U.K., Canada, Sweden, and Australia), guidance published by USAID and other U.S. Government Agencies involved in these missions, and books and articles[132] discussing reconstruction. It can also call for interviews with experts and possibly small seminars or workshops to bring them together to exchange ideas. Integrating what is known will take the form of a conceptual model or, more likely, a set of competing conceptual models that represent the alternative approaches available for the command and control of reconstruction operations.

Probably the most challenging aspect of the problem formulation phase will be the development of metrics and measurement tools that will allow comparative analysis of the alternative approaches. These are likely to involve several levels of measures of merit (measures of performance, C2 effectiveness, force effectiveness, and policy effectiveness will

---

[132] This includes academic discussions as well as reporting about particular cases and situations from the perspectives of peace operators, non-governmental organizations, international organizations, and reports from field experience and exercises (workshops, planning efforts, and simulations).

all apply) as well as specific ways to characterize the different approaches and the processes they imply.

The next phase will involve exploring those alternative approaches to identify their relative strengths, weaknesses, and limiting conditions. Assuming the initial search for knowledge was thorough and thoughtful, it is very likely that more than one viable alternative has been developed. Understanding what they have in common, how they differ, and the circumstances under which they work best would be the focus of the initial round of experimentation. This probably includes some scenario-driven simulations in which small teams apply the different approaches as well as models that compare the assumptions and processes embodied in each approach to similar situations. Agent-based models may prove useful to explore interactions between information quality, interaction structures, and processes. Success in this phase would be learning that a small number of approaches might be useful or that a single approach appears dominant over the others. Broad peer review and workshops with successful practitioners might prove useful in understanding and integrating the results from different types of experimentation.

The third phase will involve moving into more realistic environments and examining the potential utility and agility of the final candidates. These efforts will require a range of plausible scenarios or situations that sample the likely operational space where reconstruction efforts might take place. If a single approach appears dominant, then the baseline would be current doctrine and practice supported by advanced information systems. If more than one approach is being examined, the basic design should include that baseline (why invest in change if it does not look more promising than the status quo?), but the basic campaign design would be comparative. The goal is to determine which approach provides the greatest expected value and agility[133] across the range of interesting situations. This effort will no doubt involve human-in-the-loop experiments supported by constructive simulations. The "red team" or set of disruptions to be overcome will require careful development and live play in order to be credible.

Once a "best" approach is selected, or a best approach is identified for specific sets of conditions or operating environments, the campaign can move on to develop the best possible demonstration experiments, perhaps in the context of one or more ACTDs. This phase will be focused on both (1) convincing the user community that this approach was better than the existing doctrine and processes and also (2) introducing potential user populations to the mission capability package so they can help critique and improve it and become familiar enough with it to help with the later transition into the force.

COMMAND ARRANGEMENTS
FOR HOMELAND DEFENSE AND HOMELAND SECURITY

Homeland Defense (a DoD responsibility) and Homeland Security (which DoD supports) have emerged as major challenges. These are inherently not only interagency (DoD, HLS, state/local government), but international and inter-sector (private, public) problems. While cabinet-level agreements exist on how they work from a federal perspective, the serious involvement of state and local authorities, as well as a host of private organizations (e.g., telecommunications companies,

---

[133] Agility consists of robustness, resilience, responsiveness, innovation, flexibility, and adaptability in both processes and organizational arrangements.

utilities, private hospitals, the Red Cross) and the foreign governments most likely to be involved (e.g., Canada and Mexico) is also required for success. Even at the federal level, considerable potential exists for confusion over priorities and responsibilities. When state and local organizations become involved, this potential for confusion becomes exacerbated. Private entities have their own agendas and, while often willing to cooperate in an obvious national emergency, are not eager to make major investments based on hypothetical situations.

Identifying the most effective way to organize the command arrangements (how decision rights [authority, responsibility] are distributed, how information and access to information are distributed, and who interacts with whom) for Homeland Defense and Homeland Security is an important research topic with crucial practical implications. On the one hand, every emergency and threat situation is unique, so a "one size fits all" solution is unlikely. On the other hand, successful cooperation and effective synchronization will not occur by chance—they will require some preparation in the form of a coherent mission capability package. The best approach will involve: creating shared understanding of how to work together; creating the infostructure and processes; training cadres of personnel needed for effective interaction; and agreeing on how decisions will be made and implemented.

As with all significant campaigns of experimentation, this one must begin with a serious effort to assemble what is already known in order to perform Problem Formulation correctly. While lessons learned from previous efforts (e.g., the responses to 9/11, management of natural disasters such as hurricanes, management of biological threats such as SARS) will be important, most of them will do more to highlight problems with old approaches and issues to be dealt with than to suggest new ways of organizing these efforts. There is a small literature on disaster response that looks at generating better levels of interaction, distribution of information, and synergistic actions, but it is largely qualitative. Some "doctrine" has also been created by some of the organizations involved, which is another source of ideas and hypotheses. However, most of the work to identify key variables and hypothesized relationships will probably need to come from workshops that bring together practitioners from all of the types of organizations involved, theoreticians from academia (organization theory, applied psychology, communications theory, information theory, network theory, sociology, anthropology, etc.), government policy makers, and technologists. This particular campaign will need a senior steering group that seeks to ensure that all of these key perspectives are incorporated and an underlying conceptual model is developed to specify alternative ways to think about the command arrangements needed.

Because of the number of factors involved in this complicated arena and the novelty of the command arrangements needed, the initial research efforts will focus on key issues rather than attempt to formulate solutions to the whole problem. These experiments will be organized around issues such as:

- the processes of developing trust (particularly developing it rapidly while under stress);

- what types of collaboration mechanisms (e-mail, video-conferencing, shared white boards, etc.) are most useful both between organizations and between personnel in the field and operations centers;

- how information can be shared efficiently and effectively between different organizations;

- what communities of interest should be pre-established;

- how communities of interest can be rapidly created when needed; and

- how information assurance can be intelligently created and maintained to achieve appropriate risk management.

This broad series of experiments will involve some laboratory efforts, some simulations, some models (particularly of information distribution), and some experiments built into exercises and planning activities in the real world. These should be integrated by both the broad steering group looking across the set of experiments (actually mini-campaigns) and some major workshops or symposia that bring the research and development teams together and report their findings to a broad set of practitioners.

The next phase will require integrating the results of the key issue experiments into an overall, coherent approach. This will require some serious work to update and interpret the results of the earlier experiments, to craft the measures of merit by which the quality of the approach will be assessed, and to generate a series of settings for assessment that sample the space of likely and important situations intelligently.

The mission capability package needed here must stress capability and must be examined from the perspectives of all of the relevant actors, governments at all levels, industry, private entities, and relevant foreign governments. Realistically, this can only be done by creating large-scale experimentation venues within which specific experiments are embedded, rather like the Multi-National Experiment series being conducted by Joint Forces Command. These experimentation venues will provide the opportunity to involve the range of relevant actors; however, the individual experiments conducted within them must be crafted carefully to ensure that (1) they provide proper opportunities for data collection and assessment and (2) they do not interfere with one another. Over time, these experiments can move from discovery (what works, why), through hypothesis testing (this tool is better than that tool, this process works best under these circumstances), to demonstration (demonstrating the value of the selected approaches and tools to those who must rely on them). That process must be carefully controlled to avoid premature conclusions from isolate findings. The goal remains to improve knowledge over time in order to improve the quality of performance.

CONCLUSION

If one dispassionately examines the challenges of transformation and current DoD efforts at experimentation, many will conclude that our efforts toward transformation are not being adequately informed and that a greatly enhanced program of experimentation is required to close the gap between what we need to know and what we do know. This book explains why simply conducting more experiments will not provide the knowledge, understanding, and experience we need. Unless we start thinking about campaigns of experimentation before we start thinking about specific experiments and unless we develop the competency to conduct quality experiments and harvest the results in the context of these campaigns of experimentation, the gap between what we have and what we need for transformation will not be closed and our transformation efforts will remain less well-informed than they should be.

Mastering the art and science of experimentation is necessary but not sufficient for successful transformation. Well-conceived and executed campaigns of experimentation will only have marginal value if we do not "let go" of the controllable variables and allow investigations that stretch our imaginations and take us well outside our comfort zones. In 2001, DoD's NCW Report to the Congress stated that we have merely scratched the surface of what is possible. Regretfully, this is the case today as well. The problem is a set of beliefs about the nature of military operations and command and control that is a result of adaptations to past conditions. Freeing our minds to think differently is perhaps the greatest of the challenges we face.

Appendix: Checklists

This appendix contains two checklists: one for conducting campaigns and one for conducting experiments. Both were designed to be used as planning and management tools over the course of either a single experiment or a campaign of experimentation. While obviously generic (and requiring tailoring to more perfectly address the concerns of any given endeavor), these checklists are intended to be used as starting points for the planning, execution, and resolution of experimentation activities.

One of the benefits of such tools is that they provide a format for structuring and articulating ideas, as well as introducing topics for discussion. Even if a given point is not relevant to a particular endeavor, it is valuable to at least *acknowledge* that fact, as well as to understand *why* it is so.

Another, more immediate benefit of the checklist tool is that it helps the planner, sponsor, or scientist simply to remember every aspect of the large and complex process that he or she is about to embark upon. This requires thought and consideration at every level, and any tool that makes the process even a little easier (or less prone to error) is valuable.

THE CAMPAIGN CHECKLIST

This list of questions addresses the overarching issues of conceptualizing, planning, and executing a series or campaign of experiments. Important concerns include: clear and broad communications and recordings of ideas, plans, and research; thorough planning before the campaign as well as the individual experiments and their stages; and flexibly accepting all types of surprises, set-backs, delays, interruptions, discoveries, and "failures" as natural and necessary aspects of experimentation work.

1. Has the need for one or more innovations or novel capabilities been articulated?

2. Have appropriate linkages been established to the relevant communities

   a. Within DoD?

   b. With interagency partners?

   c. With coalition partners?

   d. With international organizations?

   e. With non-governmental partners?

3. Is the theory underlying the experimentation explicit?

4. Has background research been carried out to determine what is known about the topic(s) of interest?

5. Have the focus and objectives of the campaign been explicitly established?

6. Has a specific conceptual model been articulated, including

a. Variables to be considered, including

- Independent variables?

- Intervening variables?

- Dependent variables?

- Possible limiting conditions?

b. Have the controllable variables been distinguished from the uncontrollable?

c. Relationships among the variables, including (as possible)

- Direction?

- Valence?

- Strength?

7. Has provision been made to maintain and update the conceptual model and the theory?

a. Is it clear who (what group of people) is responsible for these updates?

b. Has the opportunity for such updates been scheduled?

8. Have a dictionary, lexicon, and data dictionary been established? Is someone both tasked with maintaining them over time and empowered to make them authoritative within the campaign?

9. Has provision been made for a data structure (metadata and relational database) to capture the results of

individual experiments in a way that will allow their comparison and integrated analysis?

10. Is analytical expertise available to review the results of individual experiments and to conduct analyses across experiments?

11. Has a group been established to lead the campaign?

   a. Does it have appropriate linkages to the stakeholders?

   b. Does it have linkages to those who would be responsible for implementing and supporting the innovation following a successful campaign of experimentation?

   c. Does it include substantive (domain) expertise, expertise in experimentation, and relevant experience in both arenas?

   d. Are the roles clearly defined within this group, including who has what decision rights?

   e. Does it have adequate staff support?

   f. Does it have control over resources such that it can redirect or refocus the effort over time?

12. Has an outside peer review group been established that is (a) independent of the leadership group, (b) includes both domain and experimentation expertise, and (c) includes specialists knowledgeable of the stakeholders and transition targets?

13. Has responsibility for documenting the results of the campaign been clearly assigned? Properly resourced?

The campaign checklist

14. Does the campaign plan define the range or interesting contexts or environments relevant to the innovation and make provision for sampling that space?

15. Are the linkages between the different experiments and other campaign activities explicit and well understood?

16. Does the campaign plan make appropriate use of modeling given its focus and objectives?

17. Does the campaign plan make provision for the use of an appropriate variety of human subjects or decision-making models that incorporate a range of different decision styles?

18. Does the campaign plan have explicit propositions or hypotheses been defined for each individual experiment?

19. Does the campaign plan have measures explicitly developed for the variables of interest for each individual experiment? Are they comparable across experiments? Are the measures and measurement tools selected

    a. Valid?

    b. Reliable?

    c. Credible?

20. Are all experiments designed to include baselines or other standards that test the innovation against current practice or some capability requirement?

21. Are explicit data collection and data analysis plans developed for each individual experiment? Are they refined during the pre-test phase?

22. Where low control environments are employed (exercises or free play war games, for example) does the campaign plan allow for the appropriate use of statistical controls?

23. Has provision been made to capture and accumulate insights (unexpected findings or observations that were not anticipated) across the experiments in the campaign?

24. Has provision been made for replication of important findings as the campaign of experimentation unfolds?

25. Does the campaign plan make provision for (a) assessment of proof of concept based on rapid prototypes and lean instantiations and (b) development of state of the art, supportable, and sustainable instantiations for field implementation?

26. Does the campaign plan allow for:

    a. Discarding or restarting work on innovations that are not supported?

    b. Introducing new innovations or variables that were initially left out, but are discovered to be important?

27. Has an explicit balance been sought between experiments intended for discovery, hypothesis testing, and demonstration?

28. Have an appropriate number of opportunities been developed to expose the findings of individual experiments and or the campaign to outside review?

    a. Peer reviewers?

    b. Workshops?

    c. Papers presented at conferences or symposia?

    d. Materials posted on the Web?

    e. Journals?

29. Has provision been made for periodic assessment of the progress of the campaign?

    a. Fast tracking ideas that appear robust and promising?

    b. Scheduling replications or indepth examinations of confusing or ambiguous findings?

    c. Making adjustments to the funding pattern or pace as opportunities arise?

THE EXPERIMENTATION CHECKLIST

• AUTHORS

Dr. Larry Wiener, Mr. John Poirier,
Dr. Mark Mandeles, & Dr. Michael Bell

• DATE

January 7, 2004

• PURPOSE

This task list provides a decision support and management tool for experimentation sponsors, executive agents or action officers, and task managers. It is intended to support planning, preparation, and execution of experimentation within the context of an experimentation campaign. It is intended primarily as a guide to support individuals and teams to identify the analytical activities (the whats) needed to develop the experimentation framework, and a set of process mechanics (the hows) to achieve desired outcomes. It is also intended to assist senior decisionmakers in addressing trade-offs associated with an experimentation campaign over extended time horizons, involving complex environments, multiple stakeholders, competing research interests, and limited resources.

• BACKGROUND

This task list complements the *Code of Best Practice for Experimentation* (*COBPE*) developed by the Command and Control Research Program (CCRP) of the Assistant Secretary of Defense for NII. The *COBPE* provides an overarching approach to the conceptualization, design, and execution of individual experiments or experimentation campaigns, and was developed to investigate evolving operational concepts and other areas of interest. The *COBPE* recognizes that, as a practical matter, the dynamics of experimentation are influenced by many factors beyond the control of the experimenters, especially in large events with long planning and preparation lead times.

The work reported here was based on the Multi-INT Experimental Checklist developed by a team led by Annette Krygiel. That effort had developed an intelligence-related checklist based on the *COBPE*. The present work represents an effort to extend the Multi-INT Checklist to apply to defense-related experimentation efforts in general.

The experimentation checklist

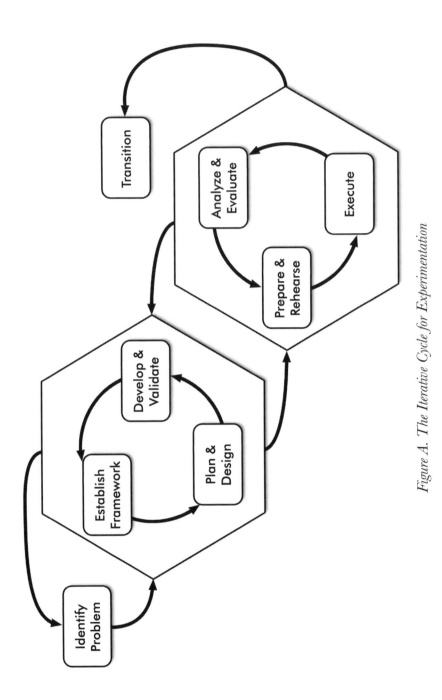

Figure A. The Iterative Cycle for Experimentation

The experimentation checklist

• FORMAT

This task list provides the elements of an iterative cycle (as shown in Figure A) combining development and assessment activities leading to the conduct of experimentation events. Critical to the success and relevance of each experiment or event is for the process to adapt to changes in priorities, environment, and stakeholder interests over the course of event preparation. There are two elements to the experiment: planning and execution. The planning process includes three critical components:

1.   Establishing an Experimentation Framework

2.   Planning and Design

3.   Development and Validation

Execution also includes three components:

1.   Preparation and Rehearsal

2.   Execution

3.   Analyses, Evaluation, and Transition

It is important to note that many of the experiment development and execution activities are cyclical and overlapping rather than sequential. Figure A provides one way to visualize this process. The process must be flexible enough to revisit components as needed. It should also be noted that as environmental conditions change and stakeholder interests are affected, options should be established for continued participation or the entry of new participants. This checklist is directed mainly at individual experiments, but the planning process described in Section 1 applies equally well to the planning of an experimentation campaign.

The experimentation checklist

1. Establishing an Experimentation Framework

    1.1 Articulate and establish research/investigative objectives.

    1.1.1   Generate an agreed upon and commonly understood statement of problem(s) to be investigated.

    1.1.2   Ensure empowered representation from the constituents of the defined community.

    1.1.3   Ensure the objectives of the investigation or experiment are measurable, and are clearly, simply, and soundly stated.

[These steps may be accomplished in a workshop or planning conference.]

    1.2 Analyze alternatives.

    1.2.1   Determine and consider alternative investigative methods to experimentation in terms of products, cost/benefit, feasibility, and risk.

    1.2.2   Determine viable approaches to investigate areas of interest, i.e., discrete events and linkages with other experiments or research activities.

[Careful articulation of information needs can facilitate obtaining data from other research activities or confirming that a gap exists.]

    1.2.3   Determine requirements for repeatability, validity, and credibility.

The experimentation checklist

1.2.4   Decide to conduct the experiment(s) including identification of needed resources and resource providers.

1.3 Identify the community of interest.

1.3.1   Identify stakeholders and potential sponsors as well as their expectations and concerns.

1.3.2   Identify "senior leadership" for the event, i.e., the decisionmaker for abort criteria.

[It is particularly important for senior leadership to be kept informed of the range of expectations and agendas represented by the various stakeholders.]

1.3.3   Define the "total" environment—research, operational, political, technical.

1.3.4   Determine range of possible effects of investigation at technical, operational, programmatic, cultural, and other levels (e.g., development of new knowledge, changed organizational structures, programmatic resources).

1.3.5   Develop a program of peer reviews, involving independent evaluations by knowledgeable personnel of the technical, programmatic, and operational merits of particular aspects of the experiment or the campaign.

1.3.6   Identify peer reviewers and enlist their participation.

1.4 Identify interests of stakeholders.

    1.4.1 Identify points of common interest and points where interests do not converge, in order to define a deconfliction process.

[The experimentation campaign planners must accommodate the concerns of those who will be affected by the experimentation process and its results.]

    1.4.2 Define the level(s) of interest for process results—national, local, service, joint, limited objective, campaign.

    1.4.3 Define, with senior leadership, the purpose and desired product types for the process, to avoid excessive focus on the output of only one event or experiment.

    1.4.4 Establish strategic, campaign level focus linking multiple events. Emphasize cumulative development of a body of knowledge that does not depend on the results of only one experiment or research activity.

1.5 Engineer flexibility into the process.

    1.5.1 Define critical decisions (rather than milestones) to allow for confirmation of sponsor and stakeholder interests, commitment to the course of action, and opportunities for the graceful exit of stakeholders who no longer wish to be part of the event as well as entry of new stakeholders.

1.5.2   Determine constraints on the experiment or on the campaign that impact iteration/ refinement cycles (see Figure A).

[Reviews should be provided to senior leadership to ensure that the ramifications of decisions and directions are recognized.]

1.6 Scope the effort.

1.6.1   Obtain general agreement on the type of experiment(s) to be conducted e.g., discovery, hypothesis testing, and demonstration,.

[In principle, it may be advisable to change the experiment type during the development process. This is a major decision requiring careful assessment and stakeholder participation.]

1.6.2   Derive the simplest experimentation approach—lab test, modeling and simulation, limited field test, etc.—to deliver desired results.

1.6.3   Document the responsibilities of each participant and obtain a common understanding and agreement among the participants.

1.6.4   Identify and execute any required contractual vehicles.

1.6.5   Determine, plan, and program to ensure availability of required prototypes, systems, materiel, databases, and infrastructure.

1.6.6   State desired outcomes, being explicit in terms of goals, metrics, applications, and relevance to type of experiment.

1.7   Establish strategic framework.

1.7.1   Identify, assess the relevance of, and pursue cooperative joint efforts with complementary research programs including those in military services/joint environments, government agencies, academe, research and development organizations, and other venues.

1.7.2   Develop contingency planning options and procedures in anticipation of unexpected events and disruptions.

1.7.3   In the case of incremental funding, ensure resource commitment and scheduling.

1.7.4   Establish tracking and management mechanisms, e.g., funding expenditures and calibrate against projected costs for the experiment, development of needed infrastructure and capabilities, task execution.

1.7.5   Develop an experiment schedule to address all phases and resource requirements.

1.7.6   Assess and account for potential impact of experimentation environment on experimentation process and outcomes.

1.7.7   Initiate actions to secure required funding, resources, and approvals.

1.8 Identify requirements.

  1.8.1 Determine minimum resources (personnel, facilities, time) to execute the experiment.

  1.8.2 Identify unique requirements for experiment conduct and design (infrastructure) to include:

  • Data collection team qualifications/ assignments.

  • Data collection team training.

  • Instrument development.

  • Consistency of involvement in experiment preparation period.

  • Selection of experimentation infrastructure (live/modeling and simulation supported).

  • Identification of all factions to be represented, e.g., Blue/Friendly, Red/ Hostile, White/Neutral, Green/Third Party, etc.

  1.8.3 Identify unique requirements for experiment participant (subject) selection and preparation to include desired expertise, training requirements, preparation timelines, language proficiency, experience sets, and Service or Combatant Command representation.

1.8.4   Determine required approvals to conduct experiments, e.g. review board approvals for use of human participants in experimentation.

1.8.5   Schedule a policy review early to identify security issues, such as need for accreditation, or policy waivers.

1.8.6   Determine requirements for, and initiate processes to obtain, waivers, such as that of security policy.

1.9   Commit resources.

1.9.1   Gain commitment of research sponsor(s) and degree of support through written agreement with participating agencies, i.e. ensure proper staff availability and secure needed resources and/or funding.

1.9.2   Assign and ensure commitment of Experiment Director/Action Officer through the research sponsor.

1.9.3   Document requirements in experiment objectives, requirements, and expected outcomes with research sponsor.

1.9.4   Ensure early identification and commitment of all experiment participants by their respective organizations—customers, process-owners, subject matter experts, prototype developers, facilities-owners, trainers, security personnel, etc.

      1.9.5 Document facility requirements and obtain agreements for their provision with hosts and supporting organizations to ensure scheduling and availability.

2.   Planning and Design

  2.1 Define role of the particular experiment within the campaign.

      2.1.1 Assess current state of knowledge in terms of what the campaign has answered so far and identify remaining unresolved issues.

      2.1.2 Inform senior leadership and confirm research priorities.

      2.1.3 Define and decide the next step in experimentation or investigation.

  2.2 Formulate experiment.

      2.2.1 Review results of research relevant to the intended experiment, including reviews of studies, experiments, customer and process-owner input, etc.

      2.2.2 Identify the products and implications of the experiment, such as residual assets, evaluation, requirements for acquisition, new business processes, etc.

      2.2.3 Ensure that the experiment incorporates appropriately mature elements, such as existing prototype versions, surrogates or mock-ups, and modeling and simulation.

2.2.4   Submit the experiment plan for peer reviews or independent assessments.

2.2.5   Develop hypotheses (IF, THEN statements) that are generated and structured to contribute to knowledge about the concept, capability, system, or process being investigated.

2.2.6   Assess prototypes, cost, risk, and schedule acceptability constraints (for experimentation).

2.3 Plan the analysis and evaluation methodology.

2.3.1   Identify control variables and determine means to treat them adequately.

[All variables should be identified by type (independent, dependent and control). Both manipulation and control are integral to hypothesis-testing experiments. The dependent variables are observed systematically under specified conditions while the factors considered to cause change—the independent variables—are varied. Other potentially relevant factors must be held constant, either empirically or through statistical manipulation. (Extracted from *COBPE*)]

2.3.2   Determine whether a documented baseline for comparison exists or the means to generate one in time for the conduct of the experiment.

2.3.3   Ensure that the evaluation method selected is relevant to the type of experiment, select appropriate metrics, and determine options to populate them through data collection.

2.3.4   Ensure participants for experiment have desired expertise, training requirements, preparation timelines, language proficiency, experience sets, and Service or Combatant Command representation.

2.3.5   Ensure processes and organizational changes are considered in the evaluation.

2.3.6   Ensure that the plans for scenario generation are sufficiently timely and that they span the conditions of interest to the sponsors.

2.3.7   Develop Analysis Plan, framework, and reporting requirements.

2.3.8   Develop data collection plan(s) for processes during the experiment and/or across the range of experiments within the campaign to ensure sufficient data to support the evaluation. Consider means to ensure that:

- Data collected reflect critical indicators.

- Quality control processes are included.

- Data collection allows for re-use and archiving through file collection and indexing.

- Standard data formats are used where possible, including date/time references.

- The evaluation plan addresses how to interpret, generalize, or scale the results of the experiment.

2.3.9 If modeling and simulation will be used to validate and expand experiment findings, then ensure that appropriate models and simulation capabilities are available, or planned.

2.4 Plan experiment and develop experimental architecture.

2.4.1 Determine and ensure understanding of applicable technical standards that must be developed and/or applied where feasible.

2.4.2 Determine if existing experimentation infrastructure in other experimentation and research venues such as Joint Forces Command (JFCOM), service laboratories, and academic institutions can be used/ leveraged.

2.4.3 Identify and describe any legacy system enhancements or needed infrastructure/tool development.

2.4.4 Identify, schedule, and ensure commitment by their owners and/or sponsors of required prototypes, systems, materiel, databases, and infrastructure, including all Government Furnished Equipment (GFE) and Government Furnished Information (GFI) by specific dates.

2.4.5   Ensure appropriate means of technical control over infrastructure development, such as configuration management processes.

2.4.6   Ensure engineering resources are available as needed to develop the end-to-end architecture to support the experiment.

2.4.7   Map appropriate elements of operational concept or problem being investigated to scenario. State research questions and framework for reporting results per the analysis plan.

2.5   Conduct facility planning.

2.5.1   Determine and schedule facilities and resources required for experiment in terms of manpower, equipment, infrastructure, etc. in accordance with the needs of each specific phase of the experiment and by specific dates.

2.5.2   Identify and resolve potential conflicts with other events supported by the experimentation facility such as training, integration and test, rehearsal, and experiment conduct.

2.5.3   Assess facility layout and determine means to minimize or avoid distractions and disruptions, such as from visitors and observers.

2.6   Develop training.

2.6.1   Determine training criteria and required standards of proficiency.

2.6.2   Plan and program training for experiment support personnel including observers, role-players, data collectors, M&S support, collection managers and others.

2.6.3   Plan and program training for experiment participants to ensure familiarity (as appropriate) with operational concepts, experiment infrastructure, experiment information exchange processes, and roles of other participants.

2.7   Conduct security planning.

2.7.1   Identify security issues and/or engineering requirements using appropriate instructions, e.g., Defense Information Technology Security Certification and Accrediation Process (DITSCAP) or National Information Assurance Certification and Accreditation Process (NIACAP).

2.7.2   Assess impact of security requirements or experiment and data availability, e.g., access by foreign nationals, "uncleared" researchers, storage of data and documents, system connectivity.

2.7.3   If needed, secure and schedule needed security resources, e.g. storage containers, destruction resources, access control resources, storage devices, etc.

2.8   Determine risk and define risk management procedures.

2.8.1   Identify types and levels of risk to experiment success, infrastructure, personnel availability, funding, schedule, and cost

2.8.2   Determine requirements and options to mitigate risk.

2.8.3   Develop risk mitigation plans for high-risk elements such as immature infrastructure components, use of surrogates, variance in software/hardware versions, long lead items, and generation of essential databases.

2.8.4   Establish guidance on documentation requirements of any design and development.

2.9 Conduct schedule planning and control.

2.9.1   Ensure that the proposed schedule is viable given the scope of the experiment, the identified risks, and the proposed funding.

2.9.2   Identify critical event dependencies and long-lead items for key activities.

2.9.3   Schedule:

- Adequate time for development and conduct of training for observers, support staff, as well as participants.

- Progress reviews for assessments of risk and progress for experiment participants, through all phases of the experiment.

- Progress reviews, peer reviews, and

critical decision reviews for programmatic goals, including one at the completion of planning phase.

- Sufficient time for testing and integration of the hardware/software/infrastructure.

2.10 Conduct transition planning.

2.10.1 Develop a transition plan to pass findings and conclusions to stakeholders.

2.10.2 Implement a communications strategy to disseminate information to participating and interested parties, e.g. form/type of briefings, Web posting, email distribution lists, etc.

2.10.3 Identify budget or Program Objective Memorandum (POM) submissions potentially affected by the experiment.

2.10.4 Identify and schedule a knowledge repository—e.g. an archive or Web site—for experimental results.

2.10.5 Ensure agreement and funding support for any residual assets either left in place or transitioned to the appropriate sponsors.

2.10.6 Estimate funding requirements for follow-on research and development activities, e.g. refinement of prototype, additional experiments, acquisitions, etc.

3. Development and Validation

3.1 Complete design and implementation plan.

3.1.1   Schedule or complete final experiment design.

3.1.2   Ensure final experiment architecture and implementation are on track.

3.1.3   Ensure development of the data collection plan is on schedule or complete, including quality control processes.

3.1.4   Ensure test and integration is proceeding on schedule, or completed.

3.1.5   Schedule and/or complete development and testing of infrastructure, support tools, required databases.

3.1.6   Ensure experiment analysis and evaluation plan are complete and all activities are on track.

3.1.7   Establish and determine application of measures of effectiveness, metrics, and/or success criteria and incorporate in the analysis and evaluation plan.

3.1.8   For experiment/research campaigns, establish iterations and entry/exit criteria.

3.1.9   If required, prepare help desk for the experiment.

3.1.10 Ensure security policies reviews are completed, accreditations obtained, or formal waivers obtained.

3.1.11 Ensure that development and implementation of operational scenarios, concepts, and script are complete or proceeding on schedule, including:

- Models and simulations.

- Training.

- Orientation and familiarization plans and materials for participants.

3.2 Execute readiness review.

    3.2.1 Schedule review of experiment preparations for the research sponsor(s).

    3.2.2 Identify a means to address problems and issues, and ensure availability of resources for corrective actions.

- Ensure data collection instruments and participant forms/questionnaires are available.

- Schedule observer interviews.

- Schedule appropriate mechanisms to capture and disseminate information such as recording of briefings.

    3.2.3 Implement help desk.

    3.2.4 Establish visitor access and control procedures.

3.2.5   Ensure data collection team and resources are ready, including a means to archive all collected data.

3.2.6   Implement process to capture Lessons Learned.

3.2.7   Implement control processes for changes to experiment environment, infrastructure, and procedures that may be caused by anomalous interruptions.

3.2.8   Describe and document rehearsal procedures for all participants and assign all responsibilities.

3.2.9   Ensure sufficient time is included in the schedule to take corrective action as needed after the rehearsal.

3.2.10 Ensure all activities in the conduct of the experiment are addressed and documented, including those of participants, observers, and visitors, in addition to equipment, systems, and infrastructure.

4.   Preparation and Rehearsal

4.1 Conduct rehearsal.

4.1.1   Include participants, observers, support personnel, and (acting) visitors engaged in the rehearsal.

4.1.2   Exercise the supporting infrastructure, including instrumentation.

4.1.3   Exercise all key activities and processes in the experiment design and plan—including data collection and analysis.

4.1.4   Ensure rehearsal includes practice vignettes and end-to-end scenarios.

4.1.5   Stress the system architecture to ensure all experiment requirements are supported.

4.1.6   Ensure a process exists to capture anomalies and unexpected disruptions and to take corrective actions in the experiment.

4.1.7   Verify that the experiment addresses the original objectives.

5.   Execution

5.1 Collect data.

5.1.1   Collect, verify, and archive data including observer notes and informal interviews.

5.1.2   Monitor quality control mechanisms for data collection.

5.2 Ensure daily communications plan is followed.

5.2.1   Schedule roundtable information exchange meetings of the data collection team and other observer groups.

5.2.2   Archive notes, data collection instruments, and other materials such as emails, voice communications logs generated during experimentation events.

5.3 Document experiment process and lessons learned.

5.3.1   Document changes (functionality) or deviations from experimentation plan in accordance with technical control plan.

5.3.2   Record down time and perceived impact on experiment in light of system or process failures such as power outages, interruptions, and loss of participants in logs and data compilation.

5.3.3   Capture and document lessons learned in experiment process.

5.3.4   Provide "hot wash" briefing and discussion for participants and capture resulting comments and observations.

6.   Analyses, Evaluation, and Transition

6.1 Provide a "quick look" report and briefing for stakeholders.

6.1.1   Report, at a minimum, the assumptions and major findings of the experiment and any initial recommendations affecting the experiment campaign or related operations.

6.1.2   Identify any unresolved issues, uncertainties, and sensitivities discovered in the experiment.

6.1.3   Provide for the widest possible circulation of the briefing and report and invite review and critique.

6.1.4   Caveat "quick reaction" results to avoid premature conclusions and recommendations.

6.2 Prepare and distribute a report of preliminary findings within a reasonably short time after the
experiment.

    6.2.1 Distinguish clearly between findings (factual
observations and data) and interpretations
of the results.

    6.2.2 Describe the effects of interruptions, disrup-
tions, anomalies, etc.

    6.2.3 Collect peer review results and incorporate
them into revised and future reports.

6.3 Prepare and publish formal reports.

    6.3.1 Adopt and implement a publication and dis-
semination plan to provide a range of
products (decision papers, summary reports,
briefings, scientific papers, articles, and
books) suited to the needs of various audiences and stakeholders.

    6.3.2 If appropriate, provide a synthesis of find-
ings and interpretations from across several
related experiments, especially those in the
experimentation campaign that includes the
present experiment.

    6.3.3 Provide recommendations for iterations of
the experiment or practical application of
results in exercises, as well as future experiments based on issues uncovered or
knowledge derived.

    6.3.4 Provide lessons learned about the experi-
mentation processes, tools used, and

infrastructure. Incorporate considerations for scaling or expanding results by further experimentation, modeling and simulation, or other means.

6.3.5   Identify and address programming and budgeting implications, including resources needed to transition results, e.g. refinement of prototypes, more experiments, implications for on-going acquisitions. Identify any budget or POM submissions affected.

6.4 Archive experiment design, data, and results for future use.

6.4.1   Collect all data records, interview transcripts, scenario descriptions, training manuals, and other artifacts immediately after execution of the experiment and preserve them in their original form.

6.4.2   Compile and archive a dictionary/glossary of the terms, constructs, and acronyms used in the experiment.

6.4.3   Compile and archive a dictionary of methodology and metrics (including definitions and scales) used in the experiment.

CONTRIBUTORS' BIOS

LARRY WIENER

Dr. Wiener has been supporting the Information Age Metrics Working Group (IAMWG) in support of the development of the Experimentation Checklist, and in the pre-publication review of other Working Group documents. He previously was an analyst in the Space and Electronic Warfare Directorate (N6) in the Office of the Chief of Naval Operations. His analytic investigations included new measures of C2 performance and their application to newly developed C2 architectures. He served as an Operations Research Analyst in Navy laboratories, in support of research and development programs. His areas of application included ocean surveillance, correlation and tracking, and ship defense. He was co-editor of a book on the analysis of urban problems. He holds a Ph.D. in Mathematics from the Catholic University of America.

MARK MANDELES

Dr. Mark D. Mandeles formed The J. de Bloch group in 1993 to examine national security and foreign policy issues. In this capacity, he has consulted for the Director of Net Assessment, the Director of Force Transformation, the Under Secretary of Defense for Policy, other Defense Department agencies, and private industry. He has published books, book chapters, encyclopedia entries, and journal articles on defense transformation, the revolution in military affairs, command and control, weapons acquisition, nuclear strategy, and ballistic missile and nuclear weapons proliferation. Dr. Mandeles earned a Ph.D. in political science from Indiana University.

JOHN POIRIER

Mr. Poirier currently is a Senior Systems Engineer and Program Manager for projects analyzing issues and defining means to realize network-centric and effects-based operational capabilities, the centerpiece of defense transformation. His activities focus on development of tools and methods to define, assess, test, and evaluate network-centric capabilities emphasizing human performance in advanced operational environments. His 23-year career includes serving in various capacities in both large and small defense firms where he has functioned in Senior Engineering, Analytical, Management, and Research roles. Mr. Poirier is also a former Navy Surface Warfare Officer and a doctoral student in Systems Engineering and Management at the George Washington University.

MICHAEL BELL

Dr. Mike Bell is a Senior Scientist at the Naval Research Laboratory in Washington, DC. He is currently assigned to the Office of the Chief of Naval Operations, Space, Information Warfare, Command and Control Directorate (N71), where he is the lead for analysis, modeling, and simulation in the Architecture and Standards Branch. His previous assignments have included tours as a staff scientist with the Chief of Naval Operations Strategic Studies Group and as a program manager at the Office of Naval Research. His recent work has focused on the structure of networks and distributed systems and the implications for network-centric military operations. Dr. Bell has published over 100 papers and reports in materials science and solid state physics and has been awarded three U.S. patents. He holds a Ph.D. in Physics from Brown University.

BIBLIOGRAPHY

"Disruptive Innovation and Retail Financial Services." BAI, Innosight. 2001. http://www.bai.org/disruption/disruptive.pdf (July 2004)

"Network Centric Warfare Report to Congress." U.S. Department of Defense. 2001. http://www.dodccrp.org/research/ncw/ncw_report/report/ncw_cover.html (July 2004)

"Transforming America's Defense: Network Centric Operations/Warfare." Pamphlet produced by the Office of the Assistant Secretary of Defense for Networks and Information Integration. 2004.

Alberts, David S., John J. Garstka, Richard E. Hayes, and David T. Signori. *Understanding Information Age Warfare*. Washington, DC: CCRP Publication Series. 2001.

Alberts, David S. and Richard E. Hayes. *Command Arrangements for Peace Operations*. Washington, DC: CCRP Publication Series. 1995.

Alberts, David S. and Richard E. Hayes. *Power to the Edge: Command...Control...In the Information Age*. Washington, DC: CCRP Publication Series. 2003.

Alberts, David S. *Information Age Transformation*. Washington, DC: CCRP Publication Series. 2002.

Alberts, David S. *The Unintended Consequences of Information Age Technologies*. Washington, DC: CCRP Publication Series. 1996.

Alberts, David S., John J. Gartska, and Frederick P. Stein. *Network Centric Warfare: Developing and Leveraging Information Superiority.* Washington DC: CCRP Publication Series. 1999.

Alberts, David S., Richard E. Hayes, Dennis K. Leedom, John E. Kirzl, and Daniel T. Maxwell. *Code of Best Practice for Experimentation.* Washington, DC: CCRP Publication Series. 2002.

American Heritage Dictionary. Third Edition. New York, NY: Houghton Mifflin Company. 1993.

Booz Allen Hamilton. NCO Conceptual Framework: Special Operations Forces Case Study. Case study directed by the Office of Force Transformation. Vienna, VA: EBR. April 2004. Available at: http://oft.ccrp050.biz/docs/NCO/workshop-4/workshop-4-navy-spec-warfare-bah

Burke, James. *Connections.* Little Brown & Co. 1978.

Campbell, Donald T. and M. Jean Russo. *Social Experimentation.* New York, NY: Sage Publications. 1998.

Cebrowski, VADM Arthur K. "What is transformation?" http://www.oft.osd.mil/what_is_transformation.cfm (July 2004)

Christensen, Clayton M. *The Innovator's Dilemma: When New Technologies Cause Great Firms to Fail.* Boston: Harvard Business School Press. 1997.

Corum, James. *The Roots of Blitzkrieg: Hans von Seeckt and German Military Reform.* Lawrence, KS: University Press of Kansas, 1994.

DARPA Strategic Plan. p. 4. http://www.darpa.mil/body/strategic.html (Mar 2004)

Definition of "Innovation." Applied Knowledge Research Institute. 2004. http://www.akri.org/cognition/inno.htm (July 2004)

Dubik, MG James. Campaign Plan 2003-2009. USJFCOM Presentation. March 2004.

Dubik, MG James. Joint Concept Development & Experimentation Campaign Plan 2003-2009 Information Briefing. Joint Forces Command. June 21, 2004.

Forbes. "Now What?" October 4, 1999.

Fourges, Pierre. "Command in Network-Centric War." Canadian Military Journal. Summer 2001.

Gladwell, Malcolm. *The Tipping Point.* Boston, MA: Back Bay Books. 2002.

Glaros, CDR Greg. "Real Options for Defense." OFT Transformation Trends. June 2003. http://www.oft.osd.mil/library/library.cfm?libcol=9 (July 2004)

Henessey, Michael A. *Magic Bullets: Historical Examples of Military Disruptive Technologies.* Royal Military College of Canada: History and War Studies. April 14, 2003. http://www.drdc-rddc.dnd.ca/newsevents/events/sympo/5.pdf. (July 2004)

Janis, Irving. *Groupthink: Psychological Studies of Policy Decision and Fiascoes.* New York, NY: Houghton Mifflin Company. 1982.

JFCOM. "What is Transformation." http://www.jfcom.mil/about/transform.html.

Keegan, John. *The Price of Admiralty: the Evolution of Naval Warfare.* New York: Penguin Books, 1988.

Keeter, Hunter, C. ed. "Giambastiani: Change in Culture Key to Joint Transformation." *Sea Power.* September 2004.

Kirzl, John, David Noble, and Dennis Leedom. "Command Performance Assessment System." Vienna, VA: EBR, Inc. 2003.

Kotter, John P. "Leading Change: Why Transformation Efforts Fail." Boston, MA: HBR OnPoint, Harvard Business School Publishing Corporation. (#4231) 2000.

Kruzins, Ed. "Factors for Network Centric Warfare: An Australian Perspective." Presented at the NEC/NCW workshop held December 17-19, 2002.

Macgregor, Douglas A. *Breaking the Phalanx*. Westport, CT: Praeger Publishers. 1997.

McNaugher, Thomas, David Johnson, and Jerry Sollinger. "Agility by a Different Measure: Creating a More Flexible U.S. Army." RAND, 2000.

Ministry of Defense. "Network Enabled Capability: an Introduction, Version 1.0." April 2004. Available at: http://www.mod.uk/issues/nec/documents.htm

Moffat, James. *Complexity Theory and Network Centric Warfare*. Washington, DC: CCRP Publication Series. 2003.

National Museum of American History. "ARPANET." http://smithsonian.yahoo.com/arpanet2.html (June 1, 2004)

NATO Code of Best Practice for C2 Assessment. Washington, DC: CCRP Publication Series. 2002.

NATO SAS-050. Terms of Reference: Exploring New Command and Control Capabilities. June 2003.

Network Centric Warfare Department of Defense Report to Congress. July 2001. http://www.dodccrp.org/research/ncw/ncw_report/report/ncw_cover.html (June 1, 2004)

Office of Force Transformation. "NCO Conceptual Framework Version 1.0." Prepared by Evidence Based Research. 2003. Available at: http://www.oft.osd.mil/library/library_files/document_353_NCO%20CF%20Version%201.0%20(FINAL).doc (June 1, 2004)

Office of Force Transformation. "NCO Conceptual Framework Version 2.0." Prepared by Evidence Based Research, Inc. Vienna, VA: 2004.

Office of Force Transformation. "Operational Sense and Respond Logistics: Coevolution of an Adaptive Enterprise Capability." Concept Document. 2004.

Office of Force Transformation. "Transformation Planning Guidance." 2003. www.oft.osd.mil/library/library_files/ document_129_Transformation_Planning_Guidance_April_200 3_1.pdf (July 2004)

Office of Force Transformation. Online Library. http://www.oft.osd.mil/library/library.cfm?libcol=2 (Oct 2004)

PA Consulting Group. "Joint US/UK Combat Operations in Operation Iraqi Freedom." Case study directed by the Office of Force Transformation. Vienna, VA: EBR. April 2004. Available at: http://oft.ccrp050.biz/docs/NCO/workshop-4/workshop-4-combat-ops-pacon

RAND. "NCO Conceptual Framework Case Study Template Illustrative Example: Adapted from RAND Air-to-Air Case Study." Prepared for the Office of Force Transformation and the Office of the Assistant Secretary of Defense (Networks and Information Integration). Vienna, VA: Evidence Based Research, Inc. 2003. Available at: http://oft.ccrp050.biz/docs/NCO/draft-air-to-air-template

RAND. Stryker Brigade Combat Team. Case study directed by the Office of Force Transformation. Vienna, VA: EBR. April 2004. Available at: http://oft.ccrp050.biz/docs/NCO/workshop-4/workshop-4-stryker-rand

Reinforce. Multinational Operations (During IRTF (L) trial of AMF (L); Amber Fox; and ISAF 3). Case study directed by the Office of Force Transformation. Vienna, VA: EBR. April 2004. Available at: http://69.59.149.228:8080/OFT/Portal/docs

Routledge Encyclopedia of Philosophy. http://www.rep.routledge.com/article/DA026#DA026P1.1 (June 1, 2004)

SAIC: Air to Ground Operations in DCX (Phase 1), Enduring Freedom and Iraqi Freedom. Case study directed by the Office of Force Transformation. Vienna, VA: EBR. April 2004. Available at: http://oft.ccrp050.biz/docs/NCO/workshop-4/workshop-4-air-ground-saic

Shadish, W.R., Cook, T.D., & Campbell, D.T. *Experimental and Quasi-Experimental Designs for Generalized Causal Inference.* Boston, MA: Houghton-Mifflin, 2002.

Siegel, Pascale. *Target Bosnia: Integrating Information Activities in Peace Operations.* Washington, DC: CCRP Publications Series. 1998.

Slater, Robert. *Jack Welch & The G.E. Way: Management Insights and Leadership Secrets of the Legendary CEO.* New York, NY: McGraw-Hill. 1998.

Spartacus Educational. "The Luddites." http://www.spartacus.schoolnet.co.uk/PRluddites.htm. (June 1, 2004)

Strategic Vision: A Paper by NATO's Strategic Commanders. Brussels, Belgium, and Suffolk, VA. March 2004.

Swedish Armed Forces Headquarters. Network Based Defense, A Smarter Way to Fight. Nov. 18, 2002. Available at: http://www.mil.se/article.php?id=7295 (June 1, 2004)

The Internet Encyclopedia of Philosophy. http://www.utm.edu/research/iep/b/bacon.htm (June 1, 2004)

The Manhattan Project Heritage Preservation Association. http://www.childrenofthemanhattanproject.org/HISTORY/ERC-1.htm (June 14, 2004)

The Royal Society of London. See: http://www.english.upenn.edu/~jlynch/Frank/Contexts/rs.html (June 1, 2004)

Thomond, P., & F. Lettice. "Disruptive Innovation Explored." Cranfield University, Cranfield, England. Presented at: 9th IPSE International Conference on Concurrent Engineering: Research and Applications (CE2002), July 2002. http://www.cranfield.ac.uk/sims/ecotech/pdfdoc/disruptive-pt.pdf

U.K. Ministry of Defense. "Network Enabled Capability: An Introduction, Version 1.0." Pamphlet. April 2004. Available at: http://www.mod.uk/issues/nec/documents.htm (June 1, 2004)

U.S. Army Training and Doctrine Command. Army Transformation Wargame 2001.U.S. Army War College: Carlisle, PA, 2001.

University of Arizona. Decision Support for U.S. Navy's Combined Task Force 50 during Enduring Freedom. Case study directed by the Office of Force Transformation. Vienna, VA: EBR. April 2004. Available at: http://oft.ccrp050.biz/docs/NCO/workshop-3/workshop-3-pdf/workshop-3-pdf-decision-support

Vance, Michael, and Diane Deacon. *Think Out of the Box*. Franklin Lakes, NJ: Career Press. 1995.

Waldholz, Michael. *Curing Cancer: The Story of the Men and Women Unlocking the Secrets of Our Deadliest Illness*. Simon & Schuster. 1997.

Webster's II New Riverside Dictionary. New York: Houghton Mifflin Company. 1996.

Webster's New World Dictionary. Third College Edition. New York: Pocket Star Books. 1995.

Webster's Third New International Dictionary. Merriam-Webster, Inc.; 3rd edition. January 2002.

Wentz, Larry. *Lessons from Bosnia: The IFOR Experience*. Washington, DC: CCRP Publication Series. 1999.

Wentz, Larry. *Lessons from Kosovo: The KFOR Experience*. Washington, DC: CCRP Publications Series. 2002.

Wheeler, Anthony J. and Ahmad R. Ganji. *Introduction to Engineering Experimentation*. Second Edition. New York, NY: Prentice Hall. 2003.

Wik, Manuel W. "Networked-Based Defense for Sweden: Latest Fashion or a Strategic Step Into the Future?" Defence Materiel Administration, FMV. 2002. Available at: http://www.kkrva.se/kkrvaht_4_2002_04.pdf (January 1, 2005)

About the Authors

Dr. David S. Alberts

Dr. Alberts is currently the Director of Research for the Office of the Assistant Secretary of Defense (Networks and Information Integration). One of his principal responsibilities is DoD's Command and Control Research Program, a program whose mission is to develop a better understanding of the implications of the Information Age for national security and command and control. Prior to this he was the Director, Advanced Concepts, Technologies, and Information Strategies (ACTIS) and Deputy Director of the Institute for National Strategic Studies at the National Defense University. Dr. Alberts was also responsible for managing the Center for Advanced Concepts and Technology (ACT) and the School of Information Warfare and Strategy (SIWS).

Dr. Alberts is credited with significant contributions to our understanding of the Information Age and its implications for national security, military command and control, and organization. These include the tenets of Network Centric Warfare, the coevolution of mission capability packages, new approaches to command and control, evolutionary acquisition of command and control systems, the nature and conduct of Information Warfare, and most recently the

"Power to the Edge" principles that currently guide efforts to provide the enabling "infostructure" of DoD transformation, and efforts to better understand edge organizations and the nature of command and control in a networked environment. These works have developed a following that extends well beyond the DoD and the Defense Industry.

His contributions to the conceptual foundations of Defense transformation are complemented by more than 25 years of pragmatic experience developing and introducing leading edge technology into private and public sector organizations. This extensive applied experience is augmented by a distinguished academic career in Computer Science and Operations Research and by Government service in senior policy and management positions.

DR. RICHARD E. HAYES

As President and founder of Evidence Based Research, Inc., Dr. Hayes specializes in multi-disciplinary analyses of command and control, intelligence, and national security issues; the identification of opportunities to improve support to decisionmakers in the defense and intelligence communities; the design and development of systems to provide that support; and the criticism, test, and evaluation of systems and procedures that provide such support. His areas of expertise include crisis management; political-military issues; research methods; experimental design; simulation and modeling; test and evaluation; military command, control, communication, and intelligence (C3I or NII); and decision-aiding systems. Since coming to Washington in 1974, Dr. Hayes has established himself as a leader in bringing the systematic use of evidence and the knowledge base of the social sciences into play in support of decisionmakers in the national security community, domestic agencies, and major corporations. He has initiated several programs of research and lines of business that achieved national attention and many others that directly influenced policy development in client organizations.

Catalog of CCRP Publications

Coalition Command and Control*
(Maurer, 1994)

Peace operations differ in significant ways from traditional combat missions. As a result of these unique
characteristics, command arrangements become far
more complex. The stress on command and control
arrangements and systems is further exacerbated by
the mission's increased political sensitivity.

The Mesh and the Net
(Libicki, 1994)

Considers the continuous revolution in information
technology as it can be applied to warfare in terms
of capturing more information (mesh) and how people and their machines can be connected (net).

Command Arrangements for
Peace Operations
(Alberts & Hayes, 1995)

By almost any measure, the U.S. experience shows
that traditional C2 concepts, approaches, and doctrine are not particularly well suited for peace
operations. This book (1) explores the reasons for
this, (2) examines alternative command arrangement
approaches, and (3) describes the attributes of effective command arrangements.

Standards: The Rough Road to the Common Byte
(Libicki, 1995)

The inability of computers to "talk" to one another is a major problem, especially for today's high technology military forces. This study by the Center for Advanced Command Concepts and Technology looks at the growing but confusing body of information technology standards. Among other problems, it discovers a persistent divergence between the perspectives of the commercial user and those of the government.

What Is Information Warfare?*
(Libicki, 1995)

Is Information Warfare a nascent, perhaps embryonic art, or simply the newest version of a time-honored feature of warfare? Is it a new form of conflict that owes its existence to the burgeoning global information infrastructure, or an old one whose origin lies in the wetware of the human brain but has been given new life by the Information Age? Is it a unified field or opportunistic assemblage?

Operations Other Than War*
(Alberts & Hayes, 1995)

This report documents the fourth in a series of workshops and roundtables organized by the INSS Center for Advanced Concepts and Technology (ACT). The workshop sought insights into the process of determining what technologies are required for OOTW. The group also examined the complexities of introducing relevant technologies and discussed general and specific OOTW technologies and devices.

Dominant Battlespace Knowledge*
(Johnson & Libicki, 1996)

The papers collected here address the most critical aspects of that problem—to wit: If the United States develops the means to acquire dominant battlespace knowledge, how might that affect the way it goes to war, the circumstances under which force can and will be used, the purposes for its employment, and the resulting alterations of the global geomilitary environment?

Interagency and Political-Military Dimensions of Peace Operations: Haiti - A Case Study
(Hayes & Wheatley, 1996)

This report documents the fifth in a series of workshops and roundtables organized by the INSS Center for Advanced Concepts and Technology (ACT). Widely regarded as an operation that "went right," Haiti offered an opportunity to explore interagency relations in an operation close to home that had high visibility and a greater degree of interagency civilian-military coordination and planning than the other operations examined to date.

The Unintended Consequences of the Information Age*
(Alberts, 1996)

The purpose of this analysis is to identify a strategy for introducing and using Information Age technologies that accomplishes two things: first, the identification and avoidance of adverse unintended consequences associated with the introduction and utilization of information technologies; and second, the ability to recognize and capitalize on unexpected opportunities.

Joint Training for Information Managers*
(Maxwell, 1996)

This book proposes new ideas about joint training for information managers over Command, Control, Communications, Computers, and Intelligence (C4I) tactical and strategic levels. It suggests a substantially new way to approach the training of future communicators, grounding its argument in the realities of the fast-moving C4I technology.

Defensive Information Warfare*
(Alberts, 1996)

This overview of defensive information warfare is the result of an effort, undertaken at the request of the Deputy Secretary of Defense, to provide background material to participants in a series of interagency meetings to explore the nature of the problem and to identify areas of potential collaboration.

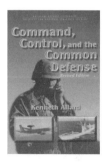

Command, Control,
and the Common Defense
(Allard, 1996)

The author provides an unparalleled basis for assessing where we are and were we must go if we are to solve the joint and combined command and control challenges facing the U.S. military as it transitions into the 21st century.

Shock & Awe:
Achieving Rapid Dominance*
(Ullman & Wade, 1996)

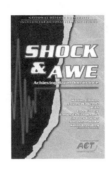

The purpose of this book is to explore alternative concepts for structuring mission capability packages around which future U. S. military forces might be configured.

Information Age Anthology:
Volume I*
(Alberts & Papp, 1997)

In this first volume, we will examine some of the broader issues of the Information Age: what the Information Age is; how it affects commerce, business, and service; what it means for the government and the military; and how it affects international actors and the international system.

Complexity, Global Politics,
and National Security*
(Alberts & Czerwinski, 1997)

The charge given by the President of the National Defense University and RAND leadership was three-fold: (1) push the envelope; (2) emphasize the policy and strategic dimensions of national defense with the implications for complexity theory; and (3) get the best talent available in academe.

Target Bosnia: Integrating Information Activities in Peace Operations*
(Siegel, 1998)

This book examines the place of PI and PSYOP in peace operations through the prism of NATO operations in Bosnia-Herzegovina.

Coping with the Bounds
(Czerwinski, 1998)

The theme of this work is that conventional, or linear, analysis alone is not sufficient to cope with today's and tomorrow's problems, just as it was not capable of solving yesterday's. Its aim is to convince us to augment our efforts with nonlinear insights, and its hope is to provide a basic understanding of what that involves.

Information Warfare and International Law*
(Greenberg, Goodman, & Soo Hoo, 1998)

The authors, members of the Project on Information Technology and International Security at Stanford University's Center for International Security and Arms Control, have surfaced and explored some profound issues that will shape the legal context within which information warfare may be waged and national information power exerted in the coming years.

Lessons From Bosnia: The IFOR Experience* (Wentz, 1998)

This book tells the story of the challenges faced and innovative actions taken by NATO and U.S. personnel to ensure that IFOR and Operation Joint Endeavor were military successes. A coherent C4ISR lessons learned story has been pieced together from firsthand experiences, interviews of key personnel, focused research, and analysis of lessons learned reports provided to the National Defense University team.

Doing Windows: Non-Traditional Military Responses to Complex Emergencies (Hayes & Sands, 1999)

This book provides the final results of a project sponsored by the Joint Warfare Analysis Center. Our primary objective in this project was to examine how military operations can support the long-term objective of achieving civil stability and durable peace in states embroiled in complex emergencies.

Network Centric Warfare (Alberts, Garstka, & Stein, 1999)

It is hoped that this book will contribute to the preparations for NCW in two ways. First, by articulating the nature of the characteristics of Network Centric Warfare. Second, by suggesting a process for developing mission capability packages designed to transform NCW concepts into operational capabilities.

Behind the Wizard's Curtain
(Krygiel, 1999)

There is still much to do and more to learn and understand about developing and fielding an effective and durable infostructure as a foundation for the 21st century. Without successfully fielding systems of systems, we will not be able to implement emerging concepts in adaptive and agile command and control, nor will we reap the potential benefits of Network Centric Warfare.

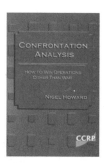

Confrontation Analysis:
How to Win Operations Other Than War
(Howard, 1999)

A peace operations campaign (or operation other than war) should be seen as a linked sequence of confrontations, in contrast to a traditional, warfighting campaign, which is a linked sequence of battles. The objective in each confrontation is to bring about certain "compliant" behavior on the part of other parties, until in the end the campaign objective is reached. This is a state of sufficient compliance to enable the military to leave the theater.

Information Campaigns
for Peace Operations
(Avruch, Narel, & Siegel, 2000)

In its broadest sense, this report asks whether the notion of struggles for control over information identifiable in situations of conflict also has relevance for situations of third-party conflict management—for peace operations.

Information Age Anthology:
Volume II*
(Alberts & Papp, 2000)

Is the Information Age bringing with it new challenges and threats, and if so, what are they? What sorts of dangers will these challenges and threats present? From where will they (and do they) come? Is information warfare a reality? This publication, Volume II of the Information Age Anthology, explores these questions and provides preliminary answers to some of them.

Information Age Anthology:
Volume III*
(Alberts & Papp, 2001)

In what ways will wars and the military that fight them be different in the Information Age than in earlier ages? What will this mean for the U.S. military? In this third volume of the Information Age Anthology, we turn finally to the task of exploring answers to these simply stated, but vexing questions that provided the impetus for the first two volumes of the Information Age Anthology.

Understanding Information Age Warfare
(Alberts, Garstka, Hayes, & Signori, 2001)

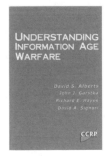

This book presents an alternative to the deterministic and linear strategies of the planning modernization that are now an artifact of the Industrial Age. The approach being advocated here begins with the premise that adaptation to the Information Age centers around the ability of an organization or an individual to utilize information.

Information Age Transformation
(Alberts, 2002)

This book is the first in a new series of CCRP books that will focus on the Information Age transformation of the Department of Defense. Accordingly, it deals with the issues associated with a very large governmental institution, a set of formidable impediments, both internal and external, and the nature of the changes being brought about by Information Age concepts and technologies.

Code of Best Practice for Experimentation
(CCRP, 2002)

Experimentation is the lynch pin in the DoD's strategy for transformation. Without a properly focused, well-balanced, rigorously designed, and expertly conducted program of experimentation, the DoD will not be able to take full advantage of the opportunities that Information Age concepts and technologies offer.

Lessons From Kosovo: The KFOR Experience
(Wentz, 2002)

Kosovo offered another unique opportunity for CCRP to conduct additional coalition C4ISR-focused research in the areas of coalition command and control, civil-military cooperation, information assurance, C4ISR interoperability, and information operations.